Rewire Your Mind

How To Change Your Mind To Live A Successful And Positive Life On Your Own Terms

Jennifer Ferguson

Copyright 2019 © Jennifer Ferguson

All rights reserved.

No part of this guide may be reproduced in any form without permission in writing from the publisher except in the case of review.

Legal & Disclaimer

The following document is reproduced below with the goal of providing information that is as accurate and reliable as possible.

This declaration is deemed fair and valid by both the American Bar Association and the Committee of Publishers Association and is legally binding throughout the United States.

Furthermore, the transmission, duplication or reproduction of any of the following work including specific information will be considered an illegal act irrespective of if it is done electronically or in print. This extends to creating a secondary or tertiary copy of the work or a recorded copy and is only allowed with an express written consent from the Publisher. All additional right reserved.

The information in the following pages is broadly considered to be a truthful and accurate account of facts, and as such any inattention, use or misuse of the information in question by the reader will render any resulting actions solely under their purview. There are no scenarios in which the publisher or the original author of this work can be in any fashion deemed liable for any hardship or damages that may befall them after undertaking information described herein.

Additionally, the information in the following pages is intended only for informational purposes and should thus be thought of as universal. As befitting its nature, it is presented without assurance regarding its prolonged validity or interim quality. Trademarks that are mentioned are done without written consent and can in no way be considered an endorsement from the trademark holder.

Table of Contents

Introduction 6

Chapter 1: Becoming Lucid 17

Chapter 2: Secrets Of The Mind You Need To Know 32

Chapter 3: The Most Complicated Object In The Known Universe 82

Chapter 4: The Body - The Unconscious Mind? 102

Chapter 5: The Expansive Environment 140

Chapter 6: Illness and Wellness 149

Chapter 7: How To Improve Brain Health With Meditation 187

Conclusion 217

Introduction

Have you ever been sleeping and dreaming a super intense and realistic dream, and suddenly in the dream you "awaken" and you realize that you are dreaming?

It's an exhilarating realization, understanding in that moment that you have the opportunity to experience the reality of your choosing - but it's fleeting. It's difficult to maintain the awakened state without leaving the dream, and you have to really focus at first to maintain your awareness. If you get good at it, through purposeful practice and repetition, you can maintain the awakened state while also allowing your mind to "reach out," if you will, as though you have one foot firmly rooted in the now, while the other is free to further engineer the events of the dream. Once you can maintain the balance of your awareness, you can remain awake as you continue to dream.

Perhaps the first time you awaken in a dream, the most you get out of it is the awareness of having woken, but then you quickly go back "under," or, upon waking in the dream, you also wake from the sleep and it's over. With practice, however, you eventually become able to maintain the awareness of being awake in the dream, while staying in the dream, and you are able to react to events with a new awareness (if it's a bad dream, you may leave the situation, or alter your reaction to it, etc.… if it's a good dream, relax and enjoy it ;). As you progress, you may find yourself better able to extend the "now" moment and stay awake and aware in the dream long enough to actually begin to engineer it and to do what you most want. Synthesize your process with your awareness in waking life, and you can extend and enhance your experience all the more.

However that may apply to your dreaming experience, returning to the present via your conscious awareness and your senses help to maintain an overall balance as far as staying "awake" within your dream, and the more that you practice being lucid, the more expansive your awareness becomes.

You can engineer your dream reality and map your journey, remaining flexible to all that comes your way by simply allowing, accepting, and adapting, because in your awakened state you know that you have the ultimate weapon: choice. Your dream is yours to shape. Do you dare to believe? If so, you must be wholly accountable for yourself. You can't be both victim and master of your destiny. You must choose which path to take – victim or master, and that doesn't mean that you won't encounter situations beyond your control, but your reaction to those situations, your perceptions, and the way in which you adapt shapes your reality, your present moment, and your future.

We also now know that the events of a dream, to the brain, aren't overly different than the events of real life – your body is just temporarily paralyzed during sleep to keep you from acting out your dreams.

So what if you could do this in real life? What if you could "awaken" within the dream of your life, of your reality, and of your perception of your reality, and take the wheel? You control your reality by being the master of your thoughts, your words, your emotions, your beliefs, your intentions, your actions, and your choices.

Waking life isn't for the faint of heart. It also

requires that you take responsibility for who you are, how you think, how you act, where you are in life, where you want to go, what you can achieve, and what you believe about your self-worth. What do you EXPECT your future to look like? What do you BELIEVE about yourself and about your ability to achieve your deepest desires? What do you believe about your deepest desires? Do you have the COURAGE to honor the desires that are closest to your heart? Do you know what those are?

Resist nothing that comes your way. A ship can't just float along, perceiving itself a victim of the river. It must sail with intention. And that doesn't mean that storms won't happen, but as a vessel you must navigate THROUGH them with clear and courageous purpose, understanding that storms are a part of life. Some are unfair because we live in a world that has lost its balance. Absurdities and atrocities happen daily, everywhere, and have for centuries. Suffering never goes away, but we can learn how to adapt, how to change, and how to survive and thrive despite life's circumstances. Theories evolve and adapt and change every day, and truths that we hold as groups of people and as individuals change as new ways of thinking emerge. How can we ever know what is true? Western medicine has only recently discovered an organ that it hadn't previously encountered, neuroscience is in the infancy of beginning to understand that the brain is inherently plastic, and the space community is discovering new planets daily. The impossible becomes increasingly possible every day, with new frontiers yet to explore. Individuals and groups are awakening and dismantling systems within themselves, in their communities, and around the world, and more and more people are rising up to meet the challenges, consequences, and dreams of the future.

As a master of your vessel, you have a destination and you are on course, ready to bend, adapt, and allow whatever comes your way, because what other choice do you have? You can't control the river, but to allow, and to flow WITH circumstances is to choose the path of least resistance, and in doing so you begin to understand that though there are many destinations, the truest destination is the journey itself. Does allowing mean condoning or silence? Of course not, but within the realm of your own mind, you can calm your own waters and make bold and enlivened moves with confidence and courage. "Right now" is the ultimate destination, and awakening to that "now" is the key to becoming the master of your world within. We obviously share the world with many people, and so it's easy to debate notions of thoughts creating reality when there are so many thoughts happening everywhere all the time. But how CAN you use your thoughts to shape your reality?

What if you are already where you need to be, with all of the necessary tools already in your arsenal? What if all that you conceive of exists, and you just have to head down to the proverbial post office to pick up your package. Do you believe that it's there waiting for you? Do you believe that you are worthy? Dare you make the journey? As they say, all journeys contain a destination that is unknown to the traveler...

You never know, maybe you set out on a journey toward one dream, and end up meeting a dream along the way that you couldn't have yet fathomed; something better than anything you could have dreamed until that point. Perhaps on your journey, you slay personal demons and you fight battles in the battlefields of thought. Perhaps you brave shipwrecks, endure heartache, survive the odds, and pull swords from stone. You conquer dragons, earn your wings, walk the darkness of your underworld, and unearth your holy grail. Perhaps you make the voyage, only to reach your destination and find that the journey WAS the destination. Perhaps the ultimate destiny is that of the phoenix – the ultimate transformation; the alchemists gold.

You must make the voyage to find out.

The concept of manifestation is to consciously and intentionally shape your personal reality through your thoughts, your beliefs, your actions, your expectations, and your emotional detachment to the outcome. How do you know whether or not some unforeseen or uncontrollable circumstance is a fortune or a misfortune? Your perception will become the truth. Remaining unattached to the outcome, emotionally, allows for peace of mind, smooth sailing, and greater efficiency. When you are in a state of allowing, or essentially, the state of unattachment, you are "going with the flow," if you will, and you are therefore moving more efficiently toward your desired destination. Does this mean that you do not experience emotion? Does this mean that you condone unfair circumstances, or that you sit passively by while life happens? No! However, what this does mean, is that in every moment you are able to return to yourself to regain your personal balance, and in doing so, you let go of what you can't control and allow the now to be what it is, including your own reactions to that now. Feel the emotions that you feel, allow your truest reactions, and observe them, embrace them, and shape them as it serves you. If you are angry, does it serve you to pretend that you aren't? If you aren't releasing your emotions... where are they going? Allowing also means allowing whatever is going on in your inner world to be where it is and when it is without judgment. How can you air out a closet if you don't open the door? Trust in your journey, and trust in yourself as the architect of your destiny. Build your ship, chart your path, know the waters, and manifest.

Throughout the course of this book we will navigate the waters of the mind, the brain and the body, illness and wellness, environment, perception, reality, intention, action, and the ways in which we can live our lives on our own terms, taking into account the importance of thoughts, words, belief systems, culture, environment, biology, choice, action, and privilege, and how these factors affect the process of changing your mind.

Chapter 1: Becoming Lucid

EVOLUTION

If we want to change our minds as individuals, we must also understand our minds within the context of history, of culture, of economic class, of privilege, and ultimately within the context of the planet. As we evolve as a species, we must think bigger and in terms of our planet more than ever before. What is evolution in the 21st century? Perhaps it is - for it must be, the evolution of consciousness; the evolution of thought, intention, and action in service of the global collective, and the undoing of the systems of the past.

If you rouse yourself from your programming and you awaken to the reality of your life, your dreams, your longing, and also your responsibilities with regard to where you are on the ship that is this planet, society, culture, etc., and you succeed in changing your mind in a holistic way (in such a way that it benefits not only you but society at large), you become responsible for the application of your new knowledge to the way you live your life. Once you awaken within the dream, you must take the wheel in as many ways as you possibly can. The little things add up, and some of the littlest things are conscious thoughts, beliefs, words, and the ways in which we interact with and view ourselves.

"Until you make the unconscious conscious, it will direct your life and you will call it fate." – Carl Jung

Have you ever noticed synchronistic symbols or numbers in your life? Repeated themes, repeated dreams or recurrent situations? Repeating patterns in relationships, cycles of behavior, or themes that re-appear throughout your life? These are the details of your waking dream – the content of your unconscious mind. According to dream expert Beverly D'Urso, the trick to becoming more lucid in your dreams is to pay attention to the details. Macrocosmically, becoming lucid and expanding your awareness in waking life is also about paying attention to the details – the details that are everywhere within your lifestyle, your daily habits, your thoughts, your dreams, your beliefs, and your reality. Why do you do what you do?

THE ART OF AWARENESS

There is a notion in Yaqui thought regarding the idea of "stalking" oneself, or essentially, "watching" or observing one's behavior as it happens. It's a far more extensive concept than we have time to visit in this book, but essentially, it's about observing your reactions. Observe your habitual thoughts and routines. Observe your moods, and if your moods shift, observe what triggered that shift. Observe your dreams, and observe the symbols of your dreams. The unconscious is both literal, and metaphorical, and it works in images – think in terms of words, language, and symbols – what programs are running beneath the surface?

Lucidity, awareness, and the return to the present keeps you balanced at the wheel of your life. When you are driving a vehicle, most of it is muscle memory, because you'd repeated the action with attention and intent so often that you had coded it into your unconscious mind and body. Regardless, whenever you operate a vehicle, you must be in the present moment, correct? You must be ever-aware of what is going on. So are you the driver or the passenger in your life? Are you asleep in the backseat? Are you asleep at the wheel? What does it take to effectively take the wheel of your life with awareness and intent?

Once you wake up, take the wheel, and practice the art of awareness, you can shift gears, raise your vibration, hit the fast lane, and explore the highest highways. Perhaps you prefer the back roads and the countryside, or you'd like to park the car and walk. Either way, directing your attention back to the present moment puts you back in the driver's seat.

CLEANING THE LINK

So basically, in order to align with the great "Intent," "Spirit," "God," "Universe, "Higher Self," the one that watches the watcher, the greater "You," you must first master your awareness and clean your connecting link. What does that mean for you?

Whatever it means for you, cleaning any sort of abstract "connection to source" would seem to require the cleaning of the spirit via the journey to the bottom of the well of the soul to explore the contents of the heart, to expose the roots of the weeds, to unearth ones deepest wounds, and to bring those treasures to the light. It's the hero's journey – the archetypal dark night of the soul, the odyssey, the quest to the phoenix; the journey from caterpillar to butterfly. That journey may take a lifetime, a moment, a day, or it may be a specific event or trauma, but whatever it is, you know it once the dawn arrives and you can look back at your journey in the glory of its light. The way through the labyrinth is to trust your intuition, to act on that trust, to let go of attachment to outcome, and to understand that the destination IS journey. If you are able to be alone with yourself, wherever you find yourself, and calm your own storm, you have achieved a milestone in the quest to master your mind. Life has many destinations and many journeys. It has many storms, heartaches, and tragedies, and the point isn't to eliminate any of these things, or to find the best way to avoid them, the point is to learn how to navigate them, so that wherever you find yourself, you can be there fully, without being capsized by the anxieties of tomorrow or the sorrows of yesterday. Be in them, be with them, embrace them, but do it on your terms rather than becoming a victim of the waves of your own emotions. The journey is an experiential way of learning some of life's most esoteric lessons because no one can take your journey but you. The treasures and the secrets hidden within your journey are yours and yours alone. No one but you can unearth them, and only you can carry them to the light.

Upon your return from the journey, transformed from ash to phoenix and armed with your flowered wounds and treasures alchemized from tragedy, you exist within a new awareness, possessing the ability to manipulate that awareness at will, because you have conquered the depths of your own Hades and risen to a new awareness of what is possible. From the vantage point of this new awareness, you are now better able to operate from the driver's seat of your vessel and manipulate the contents of your reality.

Become your own partner first before you seek without. Be your own best friend, the person you come home to every night whether you come home to someone or not. Practice self-care in the same way that you would care for a partner, and honor your needs without anger or shame. Stalk your reactions – all of them, and make your awareness of yourself a habit. Here's a real-life example of some every day stalking:

A woman burned her arm on an espresso machine. The scalding water shot out and burned her whole forearm, leaving a huge blister. She was a woman who was practiced in the art of stalking, and had been for some time, and observed herself reacting to the burn as it happened. She noted, almost immediately, that her reaction to the burn was one of love and care, rather than that of annoyance and burden, which was new and quite notable for her. It was a very subtle, essentially unconscious reaction on her part, but because she was actively stalking herself as a lifestyle, she caught the reaction as it happened and was able to observe it. She noted the change in her unconscious, achieved over time, through attention to self-care, self-awareness, and a conscious effort to change her unconscious perceptions regarding herself, and subsequently, the world.

The more that you engage in the details of life via the art of stalking, observation, or whatever you want to call it, the more exponential the experience becomes until it becomes habitual.

Take classical singers, for example, or yogis. The breathing that is practiced both in singing and in yoga is a style of breathing that is essentially intended. Breathing, for singers, causes the diaphragm to expand almost seven inches, whereas regular every day breathing moves the diaphragm about an inch or two. In the beginning, the breathing is a bit of a chore and takes much intent and effort, but after a while, the breathing becomes not only easier, but habitual and eventually unconscious.

If you want to code something into your unconscious, REPEAT IT.

It seems to follow that the cleaner the "connecting link," the higher the vibration, and the clearer the intuition. Your connecting link is your own, and your process, your feelings, and your triumphs are yours. Don't worry about how other people are cleaning their links, or how clean they are, or how far along they are in the process. You can't measure your connection to your own spirit or your life's progress by any other standard than your own, and the most effective cleaning is the most authentic one. Get real about the contents of your inner well. Embrace yourself and integrate yourself into the whole that you already see for your future.

"While we dream the assemblage point moves very gently and naturally. Mental balance is nothing but the fixing of the assemblage point on one spot we're accustomed to. Dreams make that point move, and dreaming is used to control that natural movement." – Don Juan

Lucid dreaming itself is an excellent exercise in practiced awareness, and in intended creation. Macrocosmically, examine the state of your life and practice the mental habits in your daily life that you want to code into your unconscious, and extend to your future. Get in the habit of taking note of the details of your everyday life. What does your day look like? How does that mirror, microcosmically, your whole life? Take your daily life, your habits, your thoughts, the foods you eat, the people you interact with, the things you do and use it as a micro-blueprint for your life as a whole. Alter the little things, change the little habits, and it will snowball into your bigger picture. In the same way that the littlest steps of a practice (music, sports, etc.) are fundamental to the bigger picture (muscle memory, technique), so the details of your everyday life are fundamental to the creation and outcome of your life as a whole. How do you feel about your daily life? Do you wake up happy? Excited for the day? Fulfilled? Does your day bring you energy or does it steal your energy? What things do you enjoy about your day? What things do you not enjoy? What do you want to keep? What would you like to discard or rewire?

"Sorcerers live their lives in hours." Little steps have to happen until your stride becomes confidently longer, or you'll essentially remain in the same place ever dreaming of more. It's easy to get overwhelmed by the enormity of life and of the future, but if you return to where you are you can begin to chart your next move. You don't have to know your next ten moves, or your next destination, or how you will get there, just worry about your very next move, however minute that may be. Follow those little thoughts that prompt you out of your routine, or those serendipities that seem to show up with such coincidence. Dare to take that little leap, even with only the mind, to those crazy places in the margins of the possible. Maybe that thing that you thought was a "sign," IS a sign. Dare to trust your intuition enough to take action, yet remain balanced, keeping one foot in the reality of where you are while you consider how to get where you'd like to be.

We live in a society full of authorities. Doctors and health professionals that tell you about the health of YOUR body, experts in every field imaginable feeding us endless information on the "shoulds" and the "shouldn'ts," all the while uncovering new "truths" and banishing others. How can we know what is true and what isn't? Being true to oneself in modern society, and acting on that truth is oftentimes an act of bravery. Believe in yourself enough to believe in the leap of faith, while also being perfectly fine if you fall. Failing contains a magic that can only be reaped experientially.

Before anything can "be," it must first "be" within the realm of the conceptual. Before you can produce a sound with your voice, you must first "hear" that sound in the auditorium of your mind, and if you desire a reality that vibrates at a certain frequency, do what you can to attune to that frequency.

Chapter 2: Secrets Of The Mind You Need To Know

"The conscious mind determines the actions; the unconscious mind determines the reactions, and the reactions are just as important as the actions." ~E. Stanley Jones

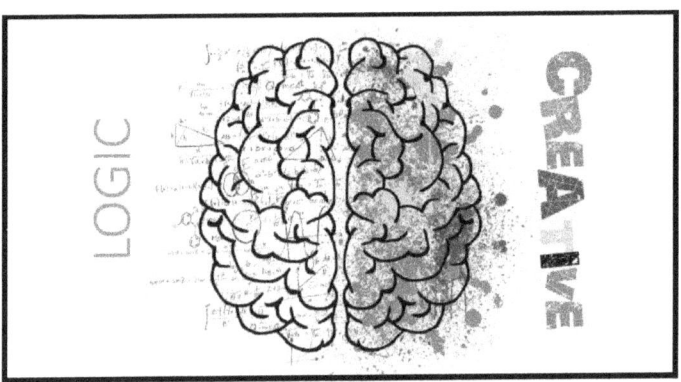

The unconscious mind plays a far larger role with regard to how we perceive reality than we may realize. Studies were conducted by Dr. John Bargh, in which participants were to meet a stranger and assess how "warm" or "cold" they perceived that person to be after having only met them for a moment. Before the subjects were introduced to the strangers, the researchers had them briefly hold a cup of either warm or iced coffee and assessed the reactions between the groups. The group that had held the warm coffee perceived the strangers to be "warmer," overall than the group that had held the cold coffee.

Why?

The unconscious is always watching, adjusting, calculating, perceiving, receiving and protecting. Much of what we consider to be conscious responses are governed largely by the unconscious mind. Have you ever been writing something while looking at something else and you accidentally write down the thing that you are looking at? By using related concepts of "warm" and "cold," researchers were able to prompt and measure an unconscious reaction to a situation across a group of people.

Were the findings also subject to the expectations of the researchers? Who can know? The mind, the brain that houses it, and the nature and contents of the universe are the greatest mysteries of science, but we are only able to observe and measure as far as our technology and tools allow. Who knows what the future of knowledge holds with regard to the power of the mind, the mysteries of the brain, and our connection to the universe around us.

All species are born into billions of years of collective memory – instincts, and human beings are no exception…. except that we have drastically changed over time, and one of our most chronic conditions as a society now, is stress. Modern stress may be a sudden fire truck siren sending your nervous system into action or a public speaking engagement, rather than a bear sighting during forest foraging or an imminent enemy. Our bodily systems are under constant attack in modern life, and stress is a chronic state for many people. When a stressful event occurs, be that an impending deadline, rush-hour traffic, or perhaps an unexpected run-in with a romantic interest, the autonomic nervous system reacts with a 'fight or flight' response. The activation of this response releases stress hormones such as cortisol and adrenaline and has rapid effects on multiple bodily systems. Heart rate and blood pressure increase, muscles tense, perspiration ensues, attention narrows, and the responses of both the digestive and immune systems are suppressed. Chronic stress also causes systemic inflammation due to the excess levels of cortisol, and cognitive functions are limited due to the narrowing of focus under duress.

Stress also adversely affects the gut due to the overflow of stress hormones that increase the permeability of the gut lining. If bacteria are detected passing through the gut wall, the immune system jumps into action, subsequently affecting the composition of the gut microbiota. A study on stress and gut bacteria in red squirrels and rat pups that had been separated from their mothers in infancy demonstrated that the greater the stress, the lower the diversity of the gut bacteria, resulting in food sensitivities, digestive issues, and an overall adverse impact on mental health.

How much is your stress worth to you?

Recent studies published by the U.S. National Library of Medicine had the following findings:

"The bidirectional communication between the central nervous system and gut microbiota, referred to as the gut-brain-axis, has been of significant interest in recent years. Increasing evidence has associated gut microbiota to both gastrointestinal and extra-gastrointestinal diseases. Dysbiosis and inflammation of the gut have been linked to causing several mental illnesses including anxiety and depression, which are prevalent in society today. Probiotics have the ability to restore normal microbial balance, and therefore have a potential role in the treatment and prevention of anxiety and depression. This review aims to discuss the development of the gut microbiota, the linkage of dysbiosis to anxiety and depression, and possible applications of probiotics to reduce symptoms. A healthy gut function has been linked to normal central nervous system (CNS) function. Hormones, neurotransmitters and immunological factors released from the gut are known to send signals to the brain either directly or via autonomic neurons.

Recently, studies have emerged focusing on variations in the microbiome and the effect on various CNS disorders, including, but not limited to anxiety, depressive disorders, schizophrenia, and autism. This review focuses on the GBA in the context of anxiety and depressive disorders. Therapeutic interventions to treat dysbiosis, or disturbance in the gut, and mitigate its effects on the GBA are only recently coming to the forefront as more is known about this unique relationship. As a result, research has been done on the use of probiotics in the treatment of anxiety and depression both as a standalone therapy and as an adjunct to commonly prescribed medications."

WAYS IN WHICH YOU CAN ALLEVIATE STRESS

- Identify the things in your life that cause both conscious and unconscious stress –

"stalk" yourself.

- Work to alleviate and balance your stress load. Obviously, things don't happen overnight, but little changes here and there in your routine will add up physically, mentally and spiritually.

- Take time for yourself – recharge in solitude, or in whatever it is that charges you – identify what brings you joy, energy, life, etc., and drink from that well often.

- Cushion your activities with time, allow your body and mind to rest, just as silence is to music – it will enliven everything else.

- Organize your priorities – where do you put your time, energy, etc.... what do you believe is most valuable? Time, rest, sleep, happiness, etc...

- Physical activity.

- Change your mind at the moment – watch a movie, go for a walk, change your mind in the moment by changing your activity and/or environment.

- Try to get enough sleep!

THOUGHTS

Returning to our discussion of the unconscious, intent, and our inherited "instincts," or memories, let's take an aside and investigate the power of words and their context within the unconscious.

Have you heard of Dr. Masaru Emoto and his water crystal experiments? Ask the internet about it. Basically, he took water and exposed it to various words, some examples include love, peace, and gratitude, and these words produced beautiful crystals once the water had been crystallized. Hateful words, as well as identities such as Adolf Hitler, produced horrid and ugly crystals. Emoto also took samples of "holy water," which produced beautiful crystals, (how did the water know it was "holy?") and water with intentions assigned to it - prayers said over it. Again, how did the water know?

Was it the expectation of the scientists?
Was it the collective memory inherent in the words and their intent and use over time?

Was it the collective assignment of meaning to the words used? How did the water know and why did it happen?
What IS collective consciousness?!

Either way, this water experiment is a testament to the effect of words on both the unconscious and conscious mind. Take negatives, for example. The unconscious is literal – if you spend your energy thinking about what you DON'T want, or what you DON'T have, your unconscious mind is going to take your thoughts literally. Perhaps you don't want to have anxiety, so rather than feeding your unconscious the idea of anxiety, ruminate rather, on what you DO want. When you are driving a car, where you direct your attention is where you end up unconsciously driving.... The more you ruminate over what you are missing, the more you essentially bathe in the energy of all that you don't have.

What about the thoughts or the beliefs that you have about yourself? How do you speak to and feel about your mainly-water self?

Words matter! Use your words to your benefit! Speak, journal, and think your reality into existence with your words on your side! Consciously changing your internal and external dialogue will extend to your unconscious world, the world around you, and effectively re-shape the linguistic content of your mind. Speak about what you want rather than what you don't. Focus on what is going right rather than what is going wrong. When you think to yourself, "I don't have any money," the unconscious generates the concept of you having no money. It doesn't understand that when you think that, you are actually thinking "I want more money." Why double your work? En-lighten-ment is just that – the raising of your vibration, and a lightening of your mental and emotional load via understanding.

The unconscious also makes associations and learns quickly. As your primary protector, the unconscious remains alert and on guard, gleaning survival lessons from every experience.

Perhaps you associate school with stress. Perhaps you had a bad experience, and your subconscious subsequently made the association between schooling and stress. Perhaps the next time you encounter school you have an immediate physical reaction – one that you can't really consciously account for, but a very immediate and visceral reaction, nonetheless. "Neurons that fire together wire together," or so it goes, and in fact, yes they do.

Have you ever been a smoker? If you have, or if you are, you'll be able to recount all the ways in which your neurons have fired together and thus wired together. Driving in the car? Smoke. Having a beer? Smoke. Just finished a big dinner? Smoke. Obviously, addiction is a factor, but the triggers are those things that are subject to the "fire together wire together" notion. Hard-wiring won't be undone in a day or two - you must painstakingly repeat the process until you achieve your desired result. Oftentimes you must first "unlearn" your current wiring, simply by being aware of it as it happens, resisting the urge to respond to the prompt, and then rewiring your neurons in the fashion of your choosing. For this reason, teachers and coaches are very specific about proper technique and form when you first begin a sport or an instrument (or whatever another undertaking).

In the beginning, your attention must be to the finer details, so as to effectively write them into your muscle memory. Once you've done that, you can relax and move on to the next task, because your body will do the rest. If, however, you code improper technique into your muscle memory, the process of correcting it is far more arduous. If your coach or your teacher tells you PLAY SLOW, it's for your own efficiency later on. The secret beneath any practice in which you must repeat until the point of mastery is to believe in the process of repetition. Many people give up before they begin to see results because they don't believe in what they cannot see. If you plant a seed, do you see it germinate? No, but yet you trust and expect that it will eventually sprout, assuming that you take proper care to nurture it. Believing in the process and acting on that belief is the most effective way to yield the desired result.

VISUALIZATION

Intentional visualization is often overlooked, if not disregarded entirely as an efficient personal practice for everyday life. Neuroscience tells us that when you imagine yourself doing something, your brain reacts in the same way to that visualization as if you were you to perform the action in real life. Obviously, then, visualization has an effect on the mind, the brain and subsequently, the body.

"As we shall see, the most remarkable feature of imagery work is that it can be accompanied by physiological changes. The beneficial physical effects of imagery would not be so surprising if we commonly thought of those mental and physical aspects as comprising two sides of a mirror that we term "body." But for three hundred years, Western medicine has separated the mind from the body. You may be surprised to learn that no other medical system in the history of the world, including Western medicine prior to the seventeenth century, makes such a distinction." - Gerald Epstein, "Healing Visualizations: Creating Health through Imagery."

There are a plethora of stories of people who have achieved the miraculous via belief and visualization. Miraculous healings, extraordinary situations and shocking achievements reached with the power of the mind have been documented around the world and throughout human history. Here is one such story on the power of belief, or the "placebo effect":

"I was particularly inspired by a recorded lecture given by Caroline Myss. She told the story of a little boy with an inoperable, malignant brain tumor. He was given only months to live, but fortunately, his oncologist suggested to his parents that he try visualizing – the only hope the physician could offer. At his parents' urging, the little boy quickly adopted a daily visualization practice. He was about eight years old and was very much interested in Star Wars. Every morning this little boy would get up and go into a meditation, during which he visualized himself as a starship captain. Every day, he would shoot at the other starships. One morning as he was visualizing, he blew up his enemy's biggest star ship. He said nothing about this to his parents. The next morning, the young lad's mother came into his bedroom and asked him to do his visualizations. He told his mother that there was no need to do them anymore. He calmly informed her that his tumor was gone. It had been blown up! Not believing him, the boy's mother pressed him to meditate, which he dutifully did. Two weeks later, the boy went to his oncologist and was given a clean bill of health. The tumor was gone. He was completely cancer free." – Lissa Rankin, M.D. "Mind Over Medicine"

Regarding the concept of visualization and manifestation, the emotions are like the wind in the sails of your vessel. Visualize a situation that you desire, and really put yourself there - experience it in your mind as viscerally as possible. Now, add emotion. Think about the energy that is contained within the realm of emotion – when you are ecstatic, what is your corresponding body state? How much momentum is contained within the purest joy or the deepest love? How much momentum lies within rage? Shame? Fear? If you find yourself in a situation where you are ruminating and overwhelmed by negative emotions, and you really want out, do your best to change course as quickly as possible to the next best option. Sometimes the mind is too tired to effectively lift itself from the undertow, but the body is there to help! The body is to mind as space is to time – the connection is inherent and holistic. The dualism of Descartes is a thing of the past, and in service of healthier communities, we must consider the elements of the human being holistically, and navigate accordingly for the future. If your mind needs help, enlist your body! Change your activity, go for a walk, listen to music, watch a movie – get out of your mind, and if the body needs help, enlist the mind!

We know, without needing to be told, that our

emotions have physiological effects. When you are angry, you FEEL it in your body. When you are sad you also feel it in your body, and the same goes for every other emotion. On some level or another, your emotional state affects your physical state, which affects your mental state, which affects your physiological state, which affects your emotional state, and so on.

DID YOU KNOW?

Over the span of the past fifty years, and more specifically, the past fifteen years, scientists have verified that the source of pain and anxiety symptoms are most often a result of what is now referred to as "cellular memory." Cellular memory is really just what we call memory, but researchers began adding the word "cellular," because although science had previously held that all memories are stored in the brain, surgeons began to find that even when they had, cumulatively and among many patients, removed every part of the brain, memory still remained. The experiences of organ transplant recipients also supports this idea:

"One famous example is Claire Sylvia, who wrote about her experience in the book "A Change of Heart." After her heart and lung transplant at Yale-New Haven Hospital in 1988, she noticed significant personality changes: she experienced strong cravings for Kentucky Fried Chicken, which as a health-conscious dancer and choreographer she would have never eaten before; she suddenly liked blues and greens rather than the bright reds and oranges she typically wore; and she became aggressive in her behavior, which was even more out of character. After some investigation, she discovered that all these new personality traits were characteristic of her donor. Dozens of similar experiences by other organ transplant recipients have been reported as well. The explanation is cellular memory." – Dr. Alexander Lloyd

In 2004, the "Dallas Morning News" ran a story called "Medical School Breakthrough," about a new study conducted at Southwestern University Medical Center in Dallas. Scientists had discovered that our experiences are recorded at the cellular level throughout our bodies, and they believed that these memories were the true source of illness and disease.

"Scientists believe these cellular memories might mean the dif-ference between a healthy life and death... Cancer can be the result of a bad cellular memory replacing a good one... This may provide one of the most powerful ways of curing illness."
– Eric Nestler, MD

As we progress through the 21st century, science and medicine are making huge leaps in understanding illness, wellness, the mind, and the body. Scientists are finding, all over the natural world, cells and organisms that record their experiences without the benefit of a brain.

Research scientist Bruce Lipton discovered, while cloning human cells, that individual muscle cells react and change based on their "perception" of an environment, and not necessarily the actual environment. (What is an "actual" environment anyway, right? Dogs can't see rainbows, color-blind people can't perceive colors that others can, etc....). Lipton's research led to the idea that human beings, as a whole, react and change based on our perceptions of, or beliefs about, our environments, and he believes that virtually every health problem originates from an errant belief on the unconscious level.

"A cellular memory that triggers fear always goes back to a wrong interpretation of the original event. The true source of my fear and stress is not the fact that Mom died; it's my belief that because Mom died, I'll never be okay again. It's not the diagnosis of cancer; it's my belief that because I've been diagnosed with cancer, my life is over. It's not the unkind thing someone did to me in and of itself; it's my belief that this unkind thing means that I am a person of inferior worth and value." – Dr. Alexander Lloyd

Dr. Lloyd goes on to reference the book, "Is This Your Child?" in which Doris Rapp writes on a cellular memory idea called the "barrel effect." Consider all of life's stress as being one big internal barrel. As long as the barrel is not full, our body can deal with new stress. Once our barrel is full, however, the littlest thing could cause a meltdown.

"Our stress barrel also includes generational memories. One could have had an idyllic childhood and a trauma-free life, but for some reason still have significant confidence issues, depression issues, health issues, or addictions. I've worked with many people who fit into that category, who later learn that a significant trauma occurred generations back—for example, a child was hit by a train and died—and no one in the family was ever the same again. These cellular memories, which are pow-erful human hard-drive viruses, are passed down like DNA. The more adrenaline released when the event happens, the stronger the cellular memory is, the more it affects you, and the more likely it is to be passed onto fu-ture generations. So the memories affecting you may not even be yours. Generational memories can explain the existence of what we started -calling "the cycle" and "breaking the cycle" a few decades back, or the behavior thought, and feeling patterns that keep repeating in certain families." – Lloyd

Karen Lawson, M.D., suggests that when we express our emotions without any attachment or judgment, we give them the freedom to flow out of our bodies and release the weight of this heavier energy. Holding onto toxic thoughts can cause a variety of problems such as high blood pressure and digestive troubles. Chronic stress can actually decrease your lifespan by shortening your telomeres (the "end caps" of DNA strands, which have a big impact on aging).

Every time you have a negative thought, your brain creates more synapses and pathways in alignment with your thought process at the time – so thinking predominantly negative thoughts will only breed more of them, while the adverse is true as well. According to many scientists, negative thinking and emotions inhibit signals from being transmitted between the central nervous system and the brain.

IMPROVE YOUR MINDSET

The sum total of all the people in the world today can be broken down into two groups, those who are always able to find success at everything they do, and those that, despite any skills or talents they may have, can never seem to get going properly. This is so because the first group has a mindset that encourages personal growth while the other does not.

The truth of the matter is that there are two very different ways people are raised when it comes to understanding ability and intelligence. Those who always seem to lack the motivation for success believe that talent and intelligence are innate and what you are born with is all you will ever get while the other, more successful, group believes that they are simply skills and that like any other skill then can be obtained via hard work and perseverance.

These two very different viewpoints, in turn, lead to dramatically different outlooks on life which eventually lead to extremely varying results. While this might seem hard to believe, for some of you anyway, heading out into the world each day with the understanding that success is possible as long as you put in the time and effort to find it will, in fact, lead to more success over time.

Known as the growth mindset, this is one thing that you can be sure all successful people have, and most of them had it instilled upon them at a very early age. At some point during childhood, everyone is either told that they succeed because they were naturally good at things or because they worked hard and never gave up.

Those who are told that they were naturally gifted often developed what is known as a fixed mindset which leads to their brains being the most active when they were receiving praise for how gifted they were. On the other hand, those that were told that their hard work was the key to their success developed what is known as a growth mindset which means their brains are most active when they are learning how to better themselves.

Fixed Mindset

- Wants to look smart or competent regardless of the reality
- Quick to avoid challenges
- Easily thwarted by obstacles
- Thinks effort is "pointless"
- Ignores feedback
- Can feel threatened by the success of others

Growth Mindset

- More interested in long-term results.
- Enjoys a challenge.
- Learns from obstacles
- Equates effort with success
- Appreciates criticism
- Finds inspiration in the success of others

To understand how the two mindsets work in action, simply remember the story of the tortoise and the hare. The hare was always told how fast he was and therefore developed a fixed mindset whereby his speed was innate and not related to his actions which meant he was free to take a nap during the race. The tortoise, on the other hand, kept a growth mindset which meant he knew that if he persevered he would succeed. This belief in himself was born out by the results of the race.

The two mindsets also manifest themselves differently when it comes to dealing with setbacks. When those who have a fixed mindset are met with a setback it directly affects how they see themselves because it shakes their belief in their innate talent. This makes it easier for them to give up on something they are struggling with as they can easily tell themselves that it is just not a talent that is in their wheelhouse. On the other hand, when a person with a growth mindset is met with a challenge they instead worry about the best way to overcome it and treat the issue as an opportunity to learn and grow.

MAXIMIZE YOUR NEUROPLASTICITY

With the consequences of having a fixed mindset so potentially disastrous, especially if you are striving to find the inner strength to empower yourself to improve your lot in life, it is important to do what you can to break these negative mental habits as quickly as possible. Luckily the human brain has the ability to constantly reshape itself throughout the course of its lifetime which means that it is never too late to shift into a growth mindset, no matter how deeply rooted the fixed mindset principles might be. New neural pathways in the brain can be formed as new thoughts are repeated time and again, and once they become well-worn paths, then new habits are formed.

While this same fact means that it will be much more difficult to change the habits that are already deeply ingrained, such as those that involve mentally keeping yourself from reaching an empowered state. While changing from a fixed to a growth mindset will be difficult, the tips outlined below will make the process more manageable than it might otherwise be.

Commit yourself to the task in front of you: Your mindset is one of the most deeply rooted patterns that your brain has gotten used to following through on over the years which means that if you ever want to empower yourself and improve it then you are really going to need to dedicte yourself to the process. It is important to keep in mind that this will be a marathon, not a sprint, and to set your expectations accordingly.

Begin with something simple: When it comes to creating new neural pathways, one of the best ways to do so is by seeing a noticeable result from an action that you consciously took in an effort to change your mindset. While a single positive choice or two a day won't generate noticeable results on their own, their cumulative effect can be substantial and that typing point can occur sooner than you might think.

Stay positive: When working to keep a growth mindset in all things, it is important to keep it up even when the going gets tough. It will likely seem like the easiest thing in the world to do while things are going well, but a fixed mindset is much more likely to manifest itself during times when roadblocks begin presenting themselves. Your fixed mindset will likely make you want to abandon all hope of forward progress when these road blocks appear.

In this case, it is important to make an effort to stop thinking of the challenges as roadblocks and start thinking of them as opportunities for you to learn and grow. Finding personal ways to meet the challenges that come your way head on without dwelling on them unnecessarily is the first step towards making a real change for the better. If you are having difficulty putting this idea into practice, consider the following:

Take the time to look for the silver lining and consider what opportunities that meeting this challenge head on will give you access to. Dealing with problems as soon as they arise will typically provide you with the opportunity to handle the issue in a simpler fashion or take actions to stop the situation from otherwise getting out of hand. Learning to appreciate this opportunity will make adopting a growth mindset much easier.

Use the challenge in question as a mirror to reflect challenges you might be having when looking to improve other facets of your life. If everything is going smoothly then it can be easy to overlook important information that could come back to bite you later if left untreated for too long. Dealing with challenges directly can, in turn, then provide you insight into what else in your life deserves a closer examination.

Think about why you feel the way you do. When face to face with roadblocks, you will find that it is often easier to approach them with a growth mindset when you take an extra moment to stop and really consider them. A rundown of the facts will often reveal that the roadblock will be much easier to overcome than it first appeared, likely because of a personal bias that made it seem much more intimidating.

Avoiding negative self-talk: Self-criticism can be beneficial if restricted to a healthy measure. However, there is a limit to the amount of negative conditioning you can subject your mind to before it starts becoming counter-productive. There is a sea of difference between, "I really need to be more physically active" than "I am a lazy blob."

Excessive self-loathing can heavily backfire because it shifts the focus from ways through which we can improve to our failures. Over a longer period, thrash negative self-speak can up your stress level, and lead to major depression. Learning to tame negative self talk is the key to overcoming challenges, feeling more confident and achieving the life of your dreams. Keep in mind that "the only thing limiting us is our belief that there are limits." Here are some tried and tested techniques for tackling the negative self-talk monster.

Throw Negative Self-Talk Into A Box: Visualize your mistakes in a tiny box the next time you find yourself exaggerating each of them. For instance, if you find yourself underperforming at a meeting or presentation, rather than thinking, it's the end of your career, try rationalizing by criticizing your choice of words. "I could have used better words or my choice of words wasn't up to the mark." This really sounds more believable than "I screwed up my career." Visualize a small box and put your poor choice of words into it. You will subconsciously diminish the problem's size, and end up feeling way more confident.

Practice Possibility Thinking: If you are constantly thinking in extremely glowing terms, you may trigger the mental lie detector which tells you that you are functioning in a surreal world. Do not force yourself to resort to extremely unreal positive thoughts. Instead, take a neutral approach when you are besieged with negative thoughts. Be more neutral in your thinking. Think about the possibilities why a certain thing could have occurred. When you feel heavy and low on energy, instead of saying, "I am a fleshy seal or fat cow" or even, "I am a goddess or diva, irrespective of how I look" try saying, "I'd be really nice if I can knock off a few pounds. It will make me feel healthier, more energetic and fitter."

In order to transition between a fixed mindset and a growth mindset, the first thing you will want to do is to consider your daily life under a microscope in the hopes of picking up on destructive habits that may be limiting your mindset without your knowledge. Once you have found these negative habits, it will be easier to act against them directly.

Look back on every day: While you are first starting out, you will likely find that you have difficultly retaining a burgeoning growth mindset as the day wears on. If this is the case, you will likely find it helpful to keep a ledger tallying all of your growth and fixed mindset thoughts throughout the day. Simply keep track of each and pay special attention to the period of time that typifies a transition from positive to negative. While this may simply be a case of the day wearing on you, it may also be related to a specific event which you are not even aware of. The only way you will know is if you map it all out in front of yourself every day for at least a month to provide you will all the data that you need to make an informed decision.

Have realistic expectations: Finally, you are going to want to keep in mind the fact that just because someone has a growth mindset doesn't mean that they are going to be happy about everything negative that befalls them. After all, everyone is simply going to have a bad day now and then; the important thing to remember is that while everyone might have an urge to give up when the going gets tough, those who have a growth mindset manage to ignore this feeling and preserve until they find the success that they have been searching for. It is important to compartmentalize these feelings rather than letting them dangerously spiral to the point that they are much more difficult to ignore than they might otherwise be. Don't let a moment of doubt turn into a day, an hour or even a minute, commit to change and find the growth mindset hiding inside you right now.

TRY AFFIRMATIONS

One great way to cut these negative thoughts off at the root is with the power of repetition. Repetition is useful when it comes to bypassing the mental filters that your fixed mindset has been allowed to create over the years when it comes to deciding how you are going to act, by default, in a given situation. Repetition is especially effective because it can allow new thoughts to slip past these filters, putting you on the path to neuroplasticity in the process.

Affirmations or mantras, positive sentences which are repeated throughout the day, are a great place to start. Affirmations are written down while mantras are repeated either aloud or in your head and both make it easier to block out any negative static that your fixed mindset has to contribute in a given situation.

When you are first stating out with this practice, it is natural to feel foolish, or as though you are wasting your time. While these thoughts are perfectly natural, if you make the mistake of acting on them, then you will be allowing your fixed mindset to assert its dominance and prevent you from making positive changes in your own life. When you are feeling especially dispirited and as though you aren't making any forward progress, it is important to power through these feelings as they are just your fixed mindset fight back. The longer you don't interact with these thoughts, the less likely they are to return.

To ensure you don't bite off more than you can chew all at once, it is recommended that you start off with an affirmation or mantra that is fairly close to your current mental comfort zone. Starting with something small will make it easier to rewire your brain in a positive direction when compared to starting with something serious right off the bat.

I Am Me: Affirmations of Self-Love

- I am a loving person who deserves to be loved in return.

- I demand respect from people around me and give respect in return.

- I am kind and caring, and I deserve to be happy.

- I am deserving of respect and admiration from my peers.

- I deserve love and kindness from those around me.

- I love the person I am just the way that I am.

- I strive to be a better person every day.

- I am worthy of love, kindness, and respect from my peers.

- I feel good about myself as an individual.

- I feel good about myself as a friend, companion, and family member.

- I am good enough for those around me.

- I am accepting and accepted by my peers.

- I am the perfect me, regardless of others' expectations.

- I set my own expectations for myself, meeting and exceeding them.

- I expect my best in any given situation, and I

give my best.

I Am Assertive: Affirmations of Self-Confidence

- I am responsible and accountable for only my own actions and reactions.

- I show my gratitude and thankfulness in new and positive ways.

- I am grateful for the positive people and energy in my life.

- I only allow positivity to affect my mental well-being.

- I only allow honesty and loyalty from those around me.

I Am Fearless: Affirmations of Courage

- I am capable of pursuing and achieving my goals.

- I always put forth my best efforts in all aspects of my life.

- I am not deterred from my short and long-term goals.

- I accept changes in my life without fear of failure or rejection.

- I emit positive energies and attract positive energies in return.

- I am brave in the face of adversity.

- I face challenges head-on and without fear.

I Am Driven: Affirmations of Ambition

- My strong work ethic is an asset to my career goals.

- I am an asset to my job and worthy of advancement to higher levels of responsibility.

- I accept constructive criticism with an open mind and heart.

- I am strong and determined to forge my own paths in life.

- I reject negative criticism in a positive manner without taking it to heart.

- I approach problem-solving with a firm belief in my abilities.

- I work hard and deserve recognition for that hard work.

- I demand no less than what my skill set is worth.

- I strive to improve my skill set every day.

- I reject negativity in the workplace, and I embrace positivity.

I Am Personable: Affirmations of Social Engagement

- I am charming and people enjoy being around me.

- I do not allow my anxiety to build a wall around me.

- I am funny and love to make people laugh.

- I do not shy away from new surroundings and new people.

- I enjoy meeting new people with like-minded spirits.

I Am Enough: Affirmations of Self-Acceptance

- I embrace the differences in my body that make me unique.

- My quirky sense of humor brings light and laughter to my life and the lives of my peers.

- I am beautiful on the inside and the outside.

- I am fortunate to be mentally and physically healthy.

- I am cautious and approach situations with

an observant and open mind.

Chapter 3: The Most Complicated Object In The Known Universe

"The human brain has 100 billion neurons, each neuron connected to ten thousand other neurons. Sitting on your shoulders is the most complicated object in the known universe." – Michio Kaku

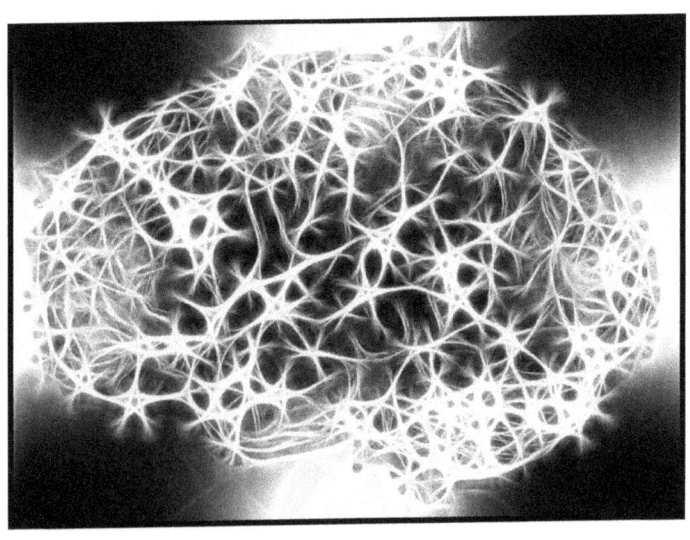

It is the brain's ability to change itself that makes it so incredible, and it should follow that if you are your body, and your body is you, then you are your brain and your brain is you. Therefore, your ability to change yourself is what makes you incredible. Great power is the ceaseless ability to adapt to circumstances and to change, and the evolutionary notion of "survival of the fittest" is just that; the fittest are those capable of the evolution necessary to move forward. Evolution waits for no one.

Until the late twentieth century, the brain was thought to have been similar to a machine – something fixed and compartmentalized by the age of eighteen, but we now know that the brain is intrinsically plastic, which means that it is literally never not plastic; whether you are one or one hundred, your brain can, does and will change itself in the moment, as you think, and as you interact with your environment.

Neuroplasticity, essentially, is the brain's ability to reorganize itself by forming new neural connections over time via repetition.

If you want to rewire your brain, repetition is paramount.

Your brain is incredible.

It's amazing.

If we were to build a computer that could simulate a human brain, it would be the size of a city block, it would require a nuclear power plant to energize it, and a lake to cool it, and yet our brains are housed in our comparatively tiny heads, and powered solely by our bodies.

Regardless of how you feel about yourself or your life, you are housing one of the most powerful and mysterious things in the universe, and not only that, you are the one powering it. You ARE your brain and your brain IS you, so you should be very proud!

THE PLASTIC BRAIN

Paul Bach-y-Rita is known as the leading visionary of neuroplasticity, and the first to propose the concept of "sensory substitution" to treat patients with neurological issues. The notion of sensory substitution is, essentially, that you feed the brain sensory information via one sense, and the hypothesis is that the brain will use and redirect that information appropriately, and, given time and repetition, more efficiently. Imagine that the information traveling around the brain is taking the "major routes," or "highways," but if an accident occurs somewhere and one of the highways becomes inoperable, the brain must re-route the information via "back roads." It must recruit, rewire, re-design, and re-plan. The more that the information travels these "back roads," the sooner these roads become the new highways.

One of the first applications of sensory substitution created by Bach-y-Rita was a chair that allowed blind people to 'see.' The trials he conducted in 1969 are now considered to be the first form of experimental evidence for neuroplasticity, and an illustration of the feasibility of sensory substitution.

Bach-y-Rita's chair had a bank of four hundred vibrating plates resting against the user's back and vibrating in connection with a camera placed above the chair facing forwards. The pattern in which the stimulation occurred enabled the user to "see," often being able to recognize an object coming toward the camera (illustrate with photos). Bach-y-Rita expected that this was neuroplasticity in action, and eventually, as technologies improved, scientists were able to prove, via brain scan, that the brain truly is plastic.

Bach-y-Rita also created a device which enabled patients with damaged vestibular nuclei (constant wobbling and the inability to remain upright) to regain their ability to remain balanced by using an electrical stimulator placed on the tongue, which reacted to a motion sensor affixed to the patient. This application enabled patients to remain balanced without the equipment after several weeks of use, indicating the strengthening of the new neural "highways" via time and repetition.

The device used by Bach-y-Rita, now sold as "Brainport," consists of a group of accelerometers attached to the patient and linked to a computer. The information is then processed and fed to a small plate positioned on the patient's tongue (due to the density of sensory receptors on the tongue). The device stimulates different areas of the tongue depending on the orientation of the accelerometers, the information is sent to the "touch" center of the brain, and is then re-routed by the brain accordingly. This stimulation allows the patient to stay balanced, and after repeated use, Bach-y-Rita discovered that the patient remained balanced for a short time after using the device.

After using the device for several weeks, the patient was completely cured, illustrating yet another application of neuroplasticity in treating neurological disorders, and also the ability of the brain to adapt to repeated stimuli. The duration was also a factor, in that the longer the patient did the exercise, the longer she was able to remain balanced after the treatment until she was fully able to return to her regular life.

This was only the seventies, at the dawn of the discovery of neuroplasticity.

It gets even crazier.

Lieutenant James Holman was a man who had become blind in his twenties as a result of chronic health troubles that began during his time in the navy. More notable, however, are the journeys and tales of his life AFTER he had become blind. James Holman, while blind, spent the majority of his life traveling the world.... alone. He trekked, by foot, across the majority of Siberia. He also mapped the Australian Outback, crossed the Indian Ocean via cargo ship, and climbed Mount Vesuvius mid-eruption, and those are only a few of his adventures. In total, he travelled 250,000 miles, which would be the equivalent of ten trips around the Equator.

He did all of this completely blind and alone.

How did he accomplish such a feat?

Holman had a few tricks to help him out – using coins as currency rather than bills, a special pocket watch that allowed him to tell time, a dictation machine to document his travels, and most magnificently, Holman had taught himself echolocation using a hickory cane, and he had re-wired his brain. It took years of determined work, and obviously, it took unwavering belief, intent, action, and repetition.

How did it work?

Imagine the sound of the cane clacking on the concrete and reaching Holman's ear. First of all, the sound vibrates the bones and membranes within his ear canal, and then the sound wave transfers its energy to a fluid in his inner ear. The fluid then sloshes over the rows of little hair cells, and bends some of them to a greater or lesser extent, depending on the sound. The hairs are connected to the dendrites of nearby nerve cells, which then fire, and transmit electrical signals through the axons and to the brain. Upon reaching the brain, the signal causes the axon to release chemicals into a nearby synapse. This arouses neurons in the auditory cortex, a patch of grey matter in the temporal lobe, where the sound is "heard."

For Holman, however, the experience doesn't end there. In order for him to consciously navigate with the sound, the signal must circulate through other patches of grey matter for further processing. Reaching that grey matter, however, requires a dive beneath the grey matter surface, and into the white matter of the brain.

Information moving within the white matter moves at speeds of up to 250 miles per hour, because the axons that move the information from one gray matter node to another are fatter than other axons, and axons are also sheathed in a fatty substance called "myelin" which insulates the axon, making the transfer of information that much more efficient. Consider the grey matter as a patchwork of chips that analyze different types of information and the white matter as cables that transmit information between those chips.

So basically, Holman's brain had to re-route the signal through the white matter to different patches of grey matter, but how do the signals know which path to take? In the same way as walking a trail through the bushes, again and again, will eventually carve out a path, so do signals in the brain follow these paths within its landscape. Really, all Holman had to do was "try" to learn echolocation, and continue to practice, and his brain did the rest of the work.

When one neuron causes another to fire over and over, the synapse between them changes in response. The axon tip of one neuron expands and begins acquiring more neurotransmitters to flood the synapse between the neurons. New axon branches may also sprout. The neuron that has been influenced, then, may lend attention back to the influencing neuron by extending more dendrite receptors toward it. Over time, the influenced neuron will respond to even mild prompts, as a pathway is forged.

Whenever Holman would click his cane, the sound waves that bounced off of various objects reached his ear at varying times. With literally years and years of repetition and practice, Holman was able to image the world around him in his mind as his brain learned to triangulate the time differences and determine the layout of the scenery surrounding him. Eventually, he could evaluate details about an object's size, shape, and texture, essentially mastering the sensory capacity of echolocation.

The brain is a remarkable thing, and even with brain scans and technology, and all of our advanced information, there remain cases of incredible feats and stories that defy logic. One such story, as referenced in Oliver Sacks' "Musicophilia," is the case of Tony Cicoria, a forty-two year old orthopedic surgeon living in a small city in upstate New York.

One afternoon Tony was attending a family gathering at a lakeside pavilion when a storm began to move in.

He went to use a pay phone outside the pavilion to make a quick call to his mother, and when he'd hung up the phone, a flash of light erupted from the phone and struck him in the face.

He'd been struck by lightning.

"I was flying forwards. Bewildered. I looked around. I saw my own body on the ground. I said to myself, 'oh shit, I'm dead.' I saw people converging on the body. I saw a woman – she had been standing waiting to use the phone right behind me – position herself over my body, give it CPR...I floated up to the stairs – my consciousness came with me. I saw my kids, had the realization that they would be okay. Then I was surrounded by a bluish-white light... an enormous feeling of well-being and peace. No emotion associated with these...pure thought, pure ecstasy. I had the perception of accelerating, being drawn up...there was speed and direction. Then, as I was saying to myself, 'this is the most glorious feeling I've ever had' – SLAM! I was back."

Upon waking, and after having had neurological

and medical examinations, it seemed that nothing was really awry with, as perplexing as that was. He'd experienced a few memory issues here and there, but otherwise, he was back to his regular life.

However...

He began experiencing an insatiable desire to listen to piano music, which was out of the ordinary for him as he hadn't really had much interest prior to his accident. He'd had a few piano lessons as a kid, but nothing notable and no interest in his adult life. He began buying recordings of the piano music of Chopin and subsequently ended up ordering all of the sheet music.

"Coincidentally" enough, one of his babysitters at the time asked him if she could store her old piano at his house, and he began to teach himself to play.

He then began to hear music in his head,

"The first time, it was a dream. I was in a tux, onstage; I was playing something I had written. I woke up, startled, and the music was still in my head." So he jumped out of bed and started trying to write it down, hardly knowing how to even notate it, and whenever he would sit down to attempt some Chopin, his own music would "come and take me over. It had a very powerful presence."

He eventually became possessed by it, obsessed; it was like the music insisted it be released. He'd wake up at four am and play until he had to go to work, and get back to it at the end of the day – it was endless. The music took over his life, and it came in "an absolute torrent of notes with no breaks, no rests between them, and he would have to give it shape and form." – Sacks

MUSIC AND THE BRAIN

"Musical activity involves nearly every region of the brain that we know of and nearly every neural subsystem." – Daniel Levitin, "This is Your Brain on Music"

How do we even begin to discuss the relationship between music and the brain? It's a romance. When music and the brain get together they're all over each other – they become one; they entrain. Music is one of the most immediate and powerful ways in which you can change your brain, your mind, your body, and your spirit because it floods your entire being in an instant. Your heart entrains to the rhythm, your body can't help but move, your breathing entrains, and may even follow the vocals (depending on whether you are hearing or listening), engaging your body as you engage in the experience. Your entire brain lights up to the stem, influencing the rest of your being. The deeper you engage, the deeper and more profound your results. We are energy, everything is energy, and music is energy intended. We all know how it feels to purge an emotion through a song or to celebrate within one. How cathartic is it to listen to music and to let it envelope you; to surrender. Is it your perception of "you," that likes music, or is it the "you" of your brain? Your body? Your mind? Your bodymind? Either way, the surrender of the human being to music, and the relationship between music and the brain is unique, and the shift in consciousness, depending on the music, is immediate, due to the entrainment of the brain and body. Music gives neuroplasticity wings, and a simple, everyday example of this is the alphabet song. Learning something by memory, dry, will take far longer than putting that same activity to a song.

Visualization of music is extremely powerful. Dare we dip into the secrets?! Let's begin with a basic example from an experiment done by neuroscientists on the power of imagery itself. The scientists had people sit at a piano and play a five-finger scale repeatedly, and they mapped the brains of the subjects while they played. The parts of the brain involved in the activity grew the more that the subjects practiced. The scientists then had another group sit at the piano and simply visualize the same experience.

The brain scans were identical.

When you listen to music, try listening to it in headphones, and actually listening to it actively, in that you put yourself at the helm of the experience. As you listen, visualize yourself AS the vocalist, or as the drummer, or the guitarist, or whatever it is – play along in your mind as the song plays into your head. Follow it with your emotion, really "be" there in it and experience it as viscerally as possible. Note your physiology afterwards.

Try it.

Not only does it feel amazing, but it activates your entire brain, body and mind, your emotions... everything, which is most likely why it feels amazing. If we had to reduce this entire book to one word, as an answer to "how to change your mind," that word would be music.

The subject of the relationship between music and the brain is vast, and far too extensive to touch on in this book, but let's briefly touch on some of the direct physiological effects of your own voice on yourself.

Chapter 4: The Body - The Unconscious Mind?

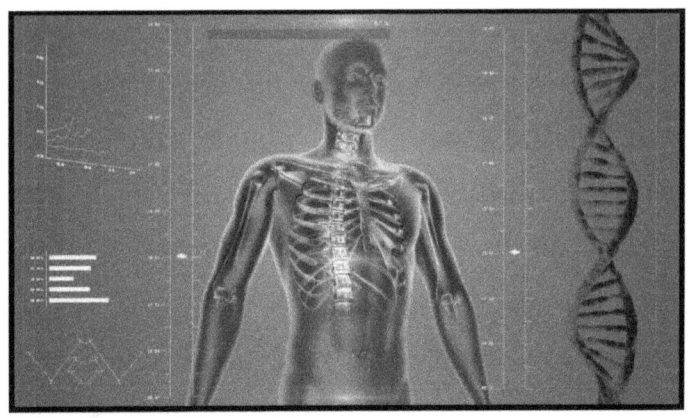

What is the collective unconscious but the "knowledge" that we've held as collectives over time? Consider the concept of Dualism, which is essentially the notion that the body and mind are separate entities. Dualism is still an infant in terms of the span of time in which humanity has existed, yet in our culture, it has remained long enough for us to forget about it and to have banked it into our collective unconscious as a "truth." Consider the body itself as the unconscious mind – makes sense, no? What are the things that you do unconsciously? It's all body function, right? Because anything conscious we call "mind" – everything else we call "body," but really, that's only because that belief is now a part of the "muscle memory" of our collective unconscious. You know those "random" thoughts that you might have, maybe you are at the grocery store and you have an overwhelming desire to buy black beans despite the fact that you rarely eat them (but maybe that's why…?), or you get a random thought to buy pickles.. but maybe not so random if you aren't one to remember to tend to your gut microbiota. Or perhaps you have a recurring body issue, and suddenly have this "strange" thought to forgive that body part. Just because the conscious mind doesn't think it doesn't mean that 'you' don't' think it – that random thought to buy pickles, or to forgive your body, or the longing for black beans... that was "you." You ARE your body. Your body knows how to get what it wants in as much as you know how to get what you want because your body is you. Our bodies are intelligent, they think, they store knowledge, hold memory, process information, change, and adapt as we interact with our environments. You are your body and your body is you. Here's an interesting exercise to try:

The next time that you speak about your body (the unconscious) "doing" something, change the focus from the "body" doing it, to "you" doing it. "You" digested your dinner, "you" washed yourself while you slept (brain), "you" activated your fight-or-flight response when you reached into your pocket for your keys and they weren't there, "you" pumped blood through your arteries this morning, and in fact, you are never not pumping blood through your arteries because it seems that you're obsessed with doing it. It's just something that you clearly enjoy; it's one of your most favorite activities, one would assume, as you literally do it twenty-four seven. What else are you obsessed with? You LOVE to use oxygen – all day long you do it, dutifully turning oxygen into carbon dioxide and churning it out. Others may wonder what's up with all the oxygen churning if they weren't also obsessed with it and doing it all day every day themselves.

Your relationship with your unconscious is your own, but there are many ways to get in touch with your body. Even the simple exercise of considering your bodily functions to be functions that you yourself are purposefully doing is enough to further ignite the dialogue between you and your body, or to be more concise, the dialogue between you and yourself. The disconnect is that these processes are unconscious and we are therefore unaware of them, but to consider those processes as "your" own and not "your body's" processes, is almost novel, as much as it seems to be common sense.

The notion that the mind and body are separate is an illustration of the repetition, over time, of the pervading concept of Dualism.

BREATHING

So what are some ways in which we can begin to master the unconscious that is our bodies? Let's begin with breathing.

If you survey a group of athletes and ask them about their experience of intentional breathing and how it relates to their performance and practice, the answers are overwhelmingly unanimous in that their relationship with their breath is a fundamental part, if not at the core of their practice. For those who participate in sports or in activities where the body is the focus, breathing is a fundamental element, and the greater the mastery of the breath and of its breadth, depth and potential, the greater the physical feats, stamina, endurance and mastery. It is your breath, after all that animates you, sustains you, and makes you what you are. You can last weeks without food, days without water, but breath exists in the realm of minutes and hours.

"Breathe," is perhaps another one-word answer to the question of how to change your mind.

Let's examine the physiology of breathing and the effects of intentional breathing on the bodymind. Intentional breathing contributes to relaxation, stress management, improvement of organ function, and overall control of psycho-physiological states.

The "pre-Botzinger" complex or "preBotC" is the name of a cluster of neurons in the brainstem discovered in nineteen ninety-one by a neurology professor at UCLA. The cluster was first discovered in mice but has since been applied to the human brain as well. In 2016, scientists Mark Krasnow and Kevin Yackle identified and studied the preBotC neurons that affected sighing in mice. Their latest findings are focusing on how these neurons affect breathing, emotional states, and alertness or arousal. The scientists have discovered the exciting correlation between breathing and changes in emotional state, not that we weren't already aware of that (taking deep breaths to calm down, for example), but it can now be observed and measured, as scientists have now located the neural circuit that causes us to calm when we breathe slowly and deeply.

Breath work, yoga, and meditation are ways in

which to combat the effect of modern stress on our bodies, as the physiological effects experienced during meditation counteract those experienced during stress. This includes sports, martial arts, and many other endeavors where focused and intentional breathing is at the core of the practice. Breathing is like a connecting link between the "body" and "mind," or, rather, the "conscious" and the "unconscious."

Breathing is unique in that it is both an unconscious function and a controllable function. Can you manipulate your endocrine system at will? Perhaps one day... but for now, we have our breath – one of our most powerful allies. Breathing communicates with the entire body and is a fundamental part of our unconscious activity. You have access to it immediately! You can immediately slow or accelerate your breathing at will, thus affecting the rest of your body. Slower breathing slows the body down, allowing for changes in a brain wave state, emotional state, perceptual state, and physiological state, and accelerating breathing excites the systemic functions and increases oxygen flow throughout the body.

Varying brain wave states can be accessed via intentional breathing.

The Delta wave state, which vibrates at between zero point five and four hertz, is the slowest brain wave state, and it is the state of deep sleep in adults. Babies between the ages of zero and two exist in the delta wave state, and it is an ideal state for unconscious programming.

Theta waves vibrate at between four and eight hertz and are the typical brainwave state during meditation. The theta state is associated with visualization, heightened intuition, and occurs in the primary stage of sleep. Children between the ages of two and six operate in the theta wave state – the realm of the imagination. This is a state in which an individual is open to suggestion (hypnosis), and will readily accept what they are told is true.

Alpha waves vibrate between eight and thirteen hertz, and the Alpha state is a state of relaxation. Children between the ages of five and eight are in the Alpha wave state, and it is associated with creativity, inspiration, ideas, and learning. The Alpha state could be considered the "gap" between the conscious and the unconscious; the portal between.

Beta waves vibrate between fourteen and twenty-nine hertz, and the Beta wave state is the typically alert waking state – the realm of conscious, analytical thinking. Children enter the Beta state after the age of eight.

Gamma waves vibrate between thirty and one

hundred hertz and are the brain waves involved in higher processing tasks and cognitive functioning. Gamma waves contribute to active learning, memory, information processing, and hyperactivity and the gamma wave state is the ideal state in which to retain information. Stimulation coupled with learning, therefore, can be a potentially valuable practice.

Notice that brain wave states operate in cycles per second, as do sounds...
We will touch on meditation again in the final chapter of this book, but for now let's head back into the archives of the unbelievable with a story of belief, intention, action, and the will to sustain the process of achieving the incredible.

"When people hear my dream of swimming in the Olympics and I tell them I want to be like Michael Phelps, people laugh ... They say, 'You are a refugee and you get £5 a day, don't waste your time and do other things.' But I believe in myself and that I can do anything. When I hear people say that, it pushes me harder. I am not just a refugee, but I am a dreamer." – Eid Al Jazairi

Do you know how sometimes your greatest obstacles become your greatest allies? So was the case for Eid Al Jazairi, a twenty-five year old Syrian refugee who fled Damascus to Britain in 2016. At the time, Eid had to cross the Mediterranean in a tiny boat, and when he arrived in Britain three years ago he wasn't able to swim. Now, he swims competitively and hopes to compete in the 2020 Olympics in Tokyo.

Al Jazairli had been training to be an accountant, and working as a visual merchandiser when the Syrian war broke out, and he arrived in Britain on a five-year visa in 2016 where he moved into a hostel. It all began one night while hanging out at a friend's place in north-east London, and he happened to catch a YouTube video of American swimmer Michael Phelps. It lit him up immediately, and after watching two hours of film, he decided that swimming was something he had to do.

Within six months he was clocking forty-three seconds for the fifty-meter freestyle, and his coaches suspect he could compete on a refugee team if the Rio 2016 innovation is repeated at the 2020 games in Tokyo.

Al Jazairli lives on an allowance of five pounds per day, and he saves on groceries to put money aside for his monthly gym membership. When he began his practice, he couldn't swim more than two or three meters, but he swims daily between six and eight am and six and eight pm. He says that swimming takes him away from everything and brings him to a world where he is untouchable.

So what was the fundamental ingredient here? Was it swimming repetitions? Was it the fact that he had a coach? His trip across the Mediterranean? His upbringing? His genes? Or was the fundamental ingredient of the decision that he made within himself to dream his dream into life? A silent knowing that he boldly acted upon.

"I'm a dreamer." – Eid Al Jazairi

FOOD

Ok, food is another one of those "one-word answers." Food is huge – food is life, it is fuel, it is medicine. That much is undeniable, wherever on the spectrum your views fall. In the West, food USED to be considered medicine, wasn't it Hippocrates that said "let thy food be thy medicine," and don't western doctors take a "Hippocratic Oath?" What does that even mean anymore? Typically, hospital food in North America is the LAST priority as far as health. In fact, food is essentially overlooked as an option for healing and is merely tended to as an obligatory practice. Here, here's your obligatory T.V. dinner for the night. Here's your "fruit" cup. It's 2019, let's not pretend that packaged food is the same as food from the ground, not that there isn't a spectrum, but it's not absurd to say that the closer to the earth that the food is, the more valuable it will be for the body, and as it moves across the spectrum of source, it can go the opposite way as well. How many "foods" now are actually essentially poisons? We can pretend we don't know but come on, we are organic beings, there's no way that inorganic substances are sustenance. Your body is just so intelligent and capable, that it fights off what you give it, around the clock. If you are eating packaged foods and fast foods around the clock, you are essentially forcing your body to be on the battlefield, around the clock. Would you give your dog a Big Mac? Probably not, right? But can you give dogs fruits, vegetables and meant? Yep! Just because we "can" eat something doesn't mean that we should. The FDA approves things that won't kill you – that doesn't mean that these things will vitalize you, and not that we don't all participate in our own ways in the variety of foods and non-foods that are out there, but understanding what is and isn't food and how it is received by your body is a beneficial practice.

That being said, how easy is it to access REAL food when you live, let's say, in small town suburbia? Or in downtown, low-income neighborhoods with fast food places and convenience stores on every corner. How about the general imbalance in the food system at large? One that is tough for the individual to navigate, without appropriate resources. So many cycles of imbalance have been born of the industrial revolution, of colonialism, of capitalism, etc.... and we are waking up to find that what we've been sold as "food," isn't, and the harmful "food" is far more affordable, available, and marketed. Our societal system is a system of scarcity in the name of "profit." Why don't companies make things that last forever? Because there's no money in that; sustainability doesn't generate profit. It's not impossible, it just isn't profitable.

Well, guess what! There's nothing profitable about exhausting our resources until we are at the point where we are not even just contemplating, but PLANNING to terraform the asteroid belt.

How much of a testament to the modern western view of health is hospital food!

Just because someone in a white coat tells you some kind of "truth" about your body doesn't mean that it overrules YOUR truths about your body. Your truths are equally valid as you participate with your doctor in service of your health. People just regurgitate what they learn – having letters after your name doesn't make anyone an authority on the experience of anyone else – it just makes them informative on the subject matter, and, depending on their approach, a potentially helpful ally in your healing experience.

What if a study came out tomorrow from the most reputable source around, saying that the majority of what doctors have learned in medical school is archaic information and is in need of an overhaul. Would you believe it? Ask yourself how much "authority" someone else needs to appear to have in order to overthrow your own authority on yourself.

"In Western medicine, the body is seen as a machine; you try to fix a broken part or take it out. In Chinese medicine, the body is seen as a garden. If the leaves are wilting or turning brown, you examine the condition of the soil; see if the plant is getting enough water and sun, or if the roots are being impinged upon. You don't just paint the leaves green!" – "Between Heaven and Earth"

The Modern Western approach to medicine is "war on disease," with doctors as generals, disease as the enemy, patients as occupied territory, and the goal is to eradicate symptoms and maximize performance.

The Eastern approach (as well as many other worldviews) is to cultivate health with doctor and patient in partnership, to improve ecological conditions, where the goal is to enhance the self-regulatory capacity of the patient. Health in this model means integrity, adaptability, and continuity. This health is sustainable.

Ayurveda

Let's explore an alternate view of food, and essentially, lifestyle in general, as the worldview of Ayurveda is an inherently holistic one. The word "Ayur" means "life," and "Veda" means science, so it is essentially a science of life and offers a body of wisdom designed to help people achieve their full human potential.

The two guiding principles of Ayurveda are that the mind and body are inextricably connected and that nothing has more power to heal and transform the body than the mind. In Ayurveda, freedom from illness depends upon the expansion of awareness, the alignment of that awareness, and the extension of that alignment and awareness to the body. Meditation and purposeful breathwork are simple examples of practices that will achieve this end. During meditation, your heart rate and breathing slow down, your body decreases the production of cortisol and adrenaline, and your increase the production of neurotransmitters that enhance well-being, including serotonin, dopamine, oxytocin, and endorphins.

Ayurveda emphasizes the practice of eating as colorful a diet as possible, and to eat your food with awareness – be in the moment WITH the food as you eat it, be aware of it in your body, be purposeful about the eating of it. Ayurvedic principles suggest that a simple way of achieving balance in your diet is to include the six "tastes," of sweet, salty, sour, pungent, bitter, and astringent in each meal. Doing so will account for all major food groups and nutrients. Including all six tastes also helps to counteract cravings and overeating.

DIETARY CHANGES

In your quest to live a life of peace and happiness, free of unnecessary stress and constant worry, look no further than what you are nourishing your body with. Everyone knows that the things we eat and drink contribute to health factors such as obesity and heart disease, but the enormous amount of other ways diet affects physical and emotional health is far less common knowledge. You are familiar with the saying, "you are what you eat," and in this chapter, we will discuss in detail the ways in which diet affects stress levels, as well as the mechanism in which specific food and drink can help or hinder your fight against chronic stress and worry.

The first thing that needs to be explained is the ways being in a constant state of stress, and rush negatively affects our dietary choices. The examples I will give are what I feel most of us can relate to. Many people will tell you that they rely almost solely on coffee to get through the day. An enormous number of people readily consume coffee every morning to start their workday and eliminate the sluggishness associated with their hectic lifestyle. There is nothing wrong with enjoying a hot cup of coffee in the morning, as it is a nice pick me up and is actually associated with numerous health benefits. The problem lies however when a person continually consumes coffee throughout their entire day.

At this point, coffee becomes an unnecessary crutch. It is very well known that the caffeine in coffee is a potent stimulant that gives a pleasant boost of energy, and is the primary reason most people consume it. You may have wondered how this actually happens, so we will discuss this mechanism. There is a neurotransmitter in the brain called adenosine. When adenosine binds to specific receptors in the brain, it acts as a type of neuronal braking system, slowing neural activity and causing you to feel sleepy.

When caffeine enters the equation, however, it actually competes with adenosine for the same receptors in the brain. Once caffeine is bound to these receptors, it actually tricks them into thinking that adenosine has attached, but since there is not actually any adenosine present, you do not experience the sleepy, tired feeling associated with it. This doesn't present much of a problem when caffeine is consumed in moderation. The issue is that when a person is constantly consuming caffeine all day every day to try and cope with a stressful schedule, adenosine never gets the chance to do its job and neuronal activity is always running in high gear.

The neurons in the brain need the chance to wind down after extended periods of activity, and when this is not allowed, the body responds by releasing the combination of stress hormones we have already discussed. Therefore, constant consumption of caffeine will in time raise your stress level, ironically making the reason you are relying on it so heavily even worse.

Another negative result that comes constantly being subjected to a stressful, hectic schedule is skipping meals and forgetting to eat. While to some this may not seem like such a bad thing, less food meaning fewer calories and weight loss, this is not actually beneficial. However, the food we consume is what gives the body the necessary vitamins and nutrients that it needs to repair itself and function at an optimal level. Make no mistake, it is important to keep a number of calories consumed every to the appropriate amount, it is equally important that you are giving your body the fuel that it needs to function properly and help you maintain your productivity.

The most detrimental dietary habit that is often the product of being constantly in a rush and stressed out is inadequate water consumption. Statistics show that a whopping seventy-five percent of Americans suffer from chronic dehydration. With such a vast majority of the population experiencing this, we really don't ever seem to notice this is occurring. This spells trouble for the health of an individual, as proper hydration is absolutely vital for virtually all bodily functions.

While I am sure you have heard the news that water makes up seventy-five percent of the body, what is more, important to point out due to the topic of this book is that water makes up an even bigger component of the human brain, eighty-five percent! How in the world can we expect our brain to function properly, keep our stress hormones at an appropriate level and allow us to manage our daily schedules when we chronically deny it the main ingredient it is made of?

Studies have shown that being even slightly dehydrated actually elevates levels of cortisol in the human body, showing a DIRECT correlation between stress and dehydration. To stress (no pun intended) even further the importance of being properly hydrated, we should discuss a few of the other serious health risks associated with not consuming enough water. Depression can actually be linked to dehydration as well, since we mentioned the brain is made up of eighty-five percent water, when dehydrated the cells in the brain actually start to malfunction, leading to feelings of grogginess, fatigue, and depression.

You are far more prone to give in to those worrisome thoughts and stressors that you are facing when you aren't properly hydrated. Furthermore, being in a constant state of dehydration is strongly associated with increased blood pressure. When the brain senses a lack of water, it instructs the pituitary gland to secrete a hormone by the name of vasopressin. This hormone causes the blood vessels to constrict, thereby increasing blood pressure throughout the body.

To wrap up the importance of maintaining proper hydration, remember that our bodies don't actually have a system in place to store water. Therefore, it is very important that the schedule you maintain from day to day includes remembering to drink water! Now that we have talked about a few of the dietary impairments that often result from always being stressed out, worried, and in a rush, we will get to some of the ways in which a proper diet can manage your stress levels, and even get into the specific food choices you can utilize.

The next food on our list that you should try to consume in your diet helps alleviate some of the primary symptoms of stress, especially if you find yourself in a situation where you are face to face with a vampire (and not in the Twilight sense of the word). I am talking about garlic. While garlic gets a bad rap because of the lingering odor that comes with it, it can actually be a powerful stress-fighting tool. Garlic is bursting with antioxidants, like most of the foods on our list. It also contains allicin, which gives our immune system the jolt it needs when dealing with a stressful life. If you can get over the pungent smell associated with it, garlic is a good idea to try and make a regular part of your diet.

While consuming the fruits and vegetables that make up most of this list is always a great idea, it's time to add another protein source to our list of foods that will help you combat chronic stress and worry. Grass-fed beef is a great dietary option to help with this. Grass-fed beef is full of more antioxidants, vitamins, and essential nutrients than regular beef while having none of the hormones and antibiotics associated with the regular beef on the market.

This meat is one of the few dietary options, besides the fatty fish, we have already discussed that can provide you with a dose of Omega-3 fatty acids. As you have probably already noticed, grass-fed beef comes with a steeper price than traditional beef, but there is a reason for this. Grass-fed beef is better for the planet, as well as the chronically stressed out people populating it. If you can, spend a little extra next time that you buy beef, grass-fed beef will help keep your body performing at an optimal level, and mitigate the stress response that we are trying to avoid.

Our dietary friend, the cow, isn't done helping us alleviate stress and worry quite yet though. The milk they produce has recently been proven to pack a stress-reducing punch as well.

While the next food on our list our list is considered a delicacy for some people, and appalling to others (there really seems to be no in between), this food actually serves its role in lowering stress levels. I am talking about oysters! These shellfish are bursting with zinc, which actually reduces the secretion of cortisol during stressful situations. By no means am I suggesting you add oysters to your diet if you find them distasteful, any well-balanced diet should be one that enjoys as well.

However, if you are already a fan of them, keep up the good work! Oysters are truly a pearl when it comes to blocking the hormones that are released when you become stressed out. Milk contains a protein called lactium, which has been proven to not only lower blood pressure but also cause a reduction of blood pressure. Scientists actually began studying the stress-reducing qualities of milk when they noticed the calming effect it had to a baby after they fed on it. While you are hopefully getting your milk from a different source than the common infant, the concentration of lactium present in a glass of milk is a great option right before bed to help ward off chronic stress, as well as improve your quality of sleep due to the concentration of magnesium also contained in milk.

The food that we will now discuss is synonymous with being in a good mood and relaxation, and not even for the actual reason it helps with this. We are talking about turkey. Turkey is usually the centerpiece of everyone's Thanksgiving meal, a time when we are surrounded by our family and loved ones, being merry and enjoying each other's company while putting the stress and worry of everyday life aside for a little. It is unclear why turkey has become such the highlight of Thanksgiving dinner, but it was an excellent choice.

Turkey contains an amino acid called

tryptophan, which has a powerful ability to increase serotonin production in our brain. This surge in serotonin results in relaxation and mood enhancement. So in the same way that we should be thankful for our lives and loved all throughout the year instead of just on Thanksgiving day, try to consume turkey on a regular basis to fight chronic stress.

So far, I hope that this chapter has given you a general idea of the types of food and beverages to include in your everyday diet. It is of the utmost importance that we do not fail to neglect healthy food choices, no matter how stressful or hectic our schedule gets. When we do this, we actually compound the stress that we experience by failing to provide the mind and body with the vitamins and nutrients that it craves to continue running smoothly. The next and final piece of advice that I would like to give the reader isn't so much a dietary option as much as it is the method in which you prepare and eat the foods we have already discussed.

How often in the fast-paced rush of your everyday life do you actually have time to prepare yourself a nutritious, homemade meal, or for that matter get to sit down and enjoy it? When we are constantly on the run, eating foods that are pre-made, pre-cooked, and nutritionally scarce is a natural result of a frantic schedule. This is especially worse if you have a family to cook for and children to consider the nutritional needs of. You may think that there is no way you can provide them with a home-cooked, delicious meal all the time, but I'd like to suggest a method.

You may or may not have heard of meal prepping in advance, but this strategy is an excellent way to provide quality nutrition to you and your family, as well as provide several other stress-relieving benefits. Meal prepping is fairly simple, you talk with your family if you have one and make a list of all the foods they would like to have for breakfast, lunch, and dinner throughout the week.

Make sure these foods not only satisfy their taste buds but also their nutritional needs as well. After your list is made, travel to your local grocery store and stock up on the food you will prepare for the entire week. If you haven't noticed, you are making ONE trip to the grocery store per week. Not only does this free up time during your week from having to run back and pick up ingredients for dinner for the night, but it also eliminates running to your local drive-thru and filling the family up on fast food.

Additionally, making one voyage to the grocery store per week alleviates stress in a hurry by saving you money. One trip means less gas you spend to and from the store, and also reduces the amount of impulse buying that comes when you spend too long staring at that box of donut holes on your way to find some nice blueberries or grass-fed beef. After you have collected all of the ingredients for your weekly nutrition, I encourage you to set aside a few hours one day per week to become your meal prep time.

During this allocated time, cook and combine all of your meals for the week at once. This may sound like a lot of work, but after you get the hang of it becomes simple. I would also suggest investing a large collection of re-sealable containers in which to store individual meals in. Getting into the habit of meal prepping for the entire week in advance gives you several advantages for leading a more stress and worry free life throughout the week. First of all, preparing all your meals in advance eliminates the guesswork of what you will be eating every day, saving your time and brainpower for other important things.

Likewise, when you are able to simply go to the fridge, grab an individually prepared meal and go, this saves you so much time and energy and allows you to ease the strain of a busy schedule. You will spend far less money if you have a game plan for all of your meals in advance than if you are routinely going to a restaurant or convenience store to pick up a quick bite to eat, as anytime you have someone else prepare food for you, it is going to cost more than doing it yourself.

Finally, meal prepping for the week in advance can greatly reduce your stress levels and peace of mind by knowing that you are providing yourself and your family with nutritious, healthy meals that will allow the body to maintain optimal performance and keep that delicate balance of stress hormones where they need to be. The art of meal prepping is a great habit of falling into if you want to make consuming foods like the ones we have discussed in this chapter a regular part of your life.

In your mission to become a more stress and worry free individual, start by taking twenty-one days to evaluate your diet and the food choices you make. During this period of habit forming, reflect back on the foods and dietary methods we have discussed in this chapter, and do your own additional research if you have any more questions about how certain foods and ingredients play a role in your ability to combat chronic stress and worry. Before you know it, you will turn to these foods without even thinking about it!

Chapter 5: The Expansive Environment

"Perception is awareness shaped by belief. Beliefs "control" perception. Rewrite beliefs and you rewrite perception. Rewrite perception and you rewrite genes and behavior...I am free to change how I respond to the world, so as I change the way I see the world I change my genetic expression. We are not victims of our

genes. We are masters of our genetics." – Bruce Lipton, Ph.D., cellular biologist

EPIGENETICS

Your genetic expression and development are shaped by your environment, and also by your reaction to your environment, and these changes can be passed down through generations.

On a social level, with regard to the environment, we are like sharks in fish tanks, in that the more expansive the environment, the more room we have to grow, the more growing we'll do and the more we will expand. This extends to the people you surround yourself with, the activities of your everyday life, your internal environment (food, thoughts, etc...) the extent of your ability to think critically, and your desire to evolve.

The more that we grow and expand, the more we positively affect ourselves, our relationships, and the world around us.

In a study by Chartrand and Bargh, researchers observed a phenomenon known as the "Chameleon Effect," which is the unconscious tendency to imitate the speech inflections, physical expressions, and body language of others in social interactions and interpersonal relationships, which researchers say can increase our likeability and make for smoother interactions. Perhaps this is the notion of "bonding" with regard to evolution and social instincts.

In the first experiment, seventy-eight individuals were asked to have a one-on-one talk with one of the researchers. Each of the researchers employed varying mannerisms - smiling, touching their faces, shaking their foot up and down, etc...

To determine whether the mannerisms had any notable effect on the interaction between the participants and the researchers, the participants were asked to participate in a second experiment in which they were sent to a room to converse with one of the researchers about a photograph. With half of the participants, the researchers maintained a neutral posture, and with the others, they mimicked their posture, movements, and mannerisms. Afterwards, the participants were asked to rate the likeability of the experimenter, and the smoothness of the interaction.

In the third experiment, researchers wanted to find out what kinds of psychological dispositions affect a person's tendency to engage in mimicry more than others. Researchers then evaluated the concept of "perspective-taking" or the degree to which people will adopt the perspectives of others.

Overall, the researchers found that the subjects whose mannerisms had been imitated had rated the researchers as more likeable, and reported having had smoother interactions with them. Additionally, the researchers found that the individuals who were more open-minded mimicked the face-rubbing gestures and foot shaking more often than others.

"Those who pay more attention mimic more, and make more friends in the process." - Chartrand

We are subject not only to the effect of others on our actions and reactions, but also our environment and the many subtleties, advertisements and subliminal messages clogging up our sensory perception. Having one foot in the awareness that you control your thoughts and desires, and another in the fact that you may be subject to environmental forces is a balanced practice. Perhaps that longing for black beans was the iron, but perhaps it was your unconscious noticing an advertisement associating black beans with happiness, success, sex, or youth. Becoming an expert in yourself allows you to expand your awareness of your patterns of thought and behavior, and recognize the various influences affecting your perceptions.

"When we become expert in something, our tastes grow more esoteric and complex. Becoming an expert in yourself, your observations will become more nuanced and complex, developing your intuition. Watch it surprise you." – Malcolm Gladwell, "Blink: The Power of Thinking Without Thinking"

"When you do not acknowledge how you really feel, you then subconsciously project onto others what you do not claim as your own." Becoming an expert in oneself, or in the art of "stalking" one's behavior, allows for a deeper awareness of motivations, patterns, and unconscious beliefs, allowing for a more expansive understanding of relationships and the ways in which we repeat unconscious behaviors until we are able to heal them by making them conscious.

What about your family environment and upbringing? How did that affect your unconscious programs? What beliefs did you acquire that you still hold? What are some that you've eradicated? What about our education system? What programs have we banked into our collective memory?

In the study "Enhanced Cognitive Flexibility in the Semi-nomadic Himba," researchers suggested that through formal education, Westerners are trained to depend on learned strategies, whereas the Himba people participate in formal education much less often, and were found to be more cognitively flexible, most likely due to a more unpredictable environment, coupled with the cognitive flexibility that comes not having been boxed in the first place. Cognitive flexibility leads to more innovative approaches and is a useful creative practice. Question everything that you know. What have you specialized in? What did you know about your practice before you were a specialist? How creative were you with regard to your practice prior to your study of it? In which ways may formal education "imprison" the creative force? How does the predictability and stability of our modern environment contribute to cognitive and creative atrophy? What can you do to spice up your environment? Your routine? Your input and output? Your thoughts and perceptions?

What about your views on the concepts of

"illness" and "wellness?"

What does health mean for you?

Chapter 6: Illness and Wellness

GLOBAL PERSPECTIVES

Have you read the book or watched the documentary "Horse Boy?" It's about a boy named Rowan who was diagnosed with extreme autism at a young age. Rowan's symptoms ranged from unexpected temper tantrums to isolation, difficulty in communication, and delayed cognitive skills. Rowan, however, found a haven in animals, and most especially horses, with whom he deeply connected. Rowan's parents had tried every kind of treatment for Rowan but hadn't had any success. Rupert, Rowan's father, who had spent some time living with the Bushmen in South Africa and had experienced their Shamanic healing powers, had researched Mongolian shamanism, and the culture of horses there, and convinced his wife to bring Rowan to Mongolia on a healing journey.

"It was just a short journey to the mountain, where nine shamans were waiting to meet us in a vast expanse of grassland. Some of the shamans were men, some women, and each was making their own preparations for the ceremony.

My doubts and fears rose to the surface again when Tulga, our English-speaking guide, introduced me to the chairman of the shamans' association of Mongolia. The man standing in front of me had a crushing grip and smelled of vodka. Had I fallen into a nest of charlatans?

'He's not good with new people,' I whispered to Tulga as he directed me to lift Rowan across to another shaman. Whispering words of comfort to my son, I passed him to the healer - as much as you can 'pass' a kicking, screaming child.

But once in her arms, to my great surprise, he went suddenly still - until the shaman's assistant passed her spiritual mistress a bottle of vodka, from which she took a hearty pull, then without warning spat the liquid all over Rowan's face and body. The result was predictable.

'Gi-raffe!' shrieked Rowan, plucking random words from thin air. 'Gotta go ho-o-ome!'

The second shaman pulled out a harp and began to play a strange tune. Then there was more vodka spitting. Rowan screamed as though being tortured for a moment before instantly calming down again.

'Let's go see some more shamans!' he shouted. We came to four shamans standing in a line, whirling, drumming, entering their trance.

Rowan gave a deep, bubbly giggle, and at that moment I knew he was okay. Actually, not just OK - I knew he had embraced the situation and was, at some level, at peace with this crazy ceremony.

The next shaman we came to whirled, sang and drummed as energetically as the others had, but whenever he came close to Rowan, his movements became quieter - gentle and slow. My son gave another of those deep-throated giggles.

'He'll be fine,' said the shaman. 'Just do this once a year for the next three years. He'll be completely healed.'

When Tulga translated this, I felt my stomach lurch. Did I dare believe it?

Then something extraordinary happened. We'd now been at the foot of the Bogd Khan for about three hours, and the light was fading. Clouds were gathering; a wind was getting up.

An eagle which had watched the entire proceedings continued to negotiate the shared territory of its branch with two ravens. The long, blue silk scarves used by the shamans snaked out from the larch limbs on the wind.

'Hey, Rowan,' I said, going over to where he'd

been playing by one of the dung and incense-smoking altars. But Rowan took no notice, his attention absorbed by the arrival of another little boy - a Mongolian child in a baseball cap and shorts.

Suddenly, without warning, Rowan ran over to the boy and started hugging him, laughing and shouting and grabbing at the newcomer's cap. Caught between annoyance and acceptance, the little boy stood stiffly but acquiescing.

This was astonishing. Rowan never took the slightest interest in other children. He was the classic autistic 'parallel player', preferring to ignore other children while playing alongside them rather than actively engage.

'Mongolian brother!' said Rowan. The boy looked surprised. A little wary, but he let Rowan hug him once more.

'Mongolian brother!' Rowan said again. 'Come on, let's go to the river!' My wife Kristin and I looked at each other in amazement.

For the first time ever, Rowan was playing with another child. Tomoo, smiling now, made a gentle cut, which Rowan parried back. At a loss for words, Kristin and I gaped as Rowan and Tomoo chased each other, laughing."

Rupert Isaacson, "The Horse Boy: A Father's Miraculous Journey To Heal His Son"

What IS the worldview of shamen on the subject of "illness" and "wellness?" What about the eastern worldview of the integrative garden? What about the Western view?

We know that the Western view of illness and health is very much "machine-based," "attack and destroy invader" mentality, and if our bodies are micro-representations of the world, this view makes perfect North American sense, doesn't it? Where does all the government money go in the United States? War? The United States in ever engaged in a war on everything.

The Shamanic view of "mental illness" is that "mental illness" signals the birth of a healer. In the Shamanic view, Rowan was a Shaman.

"From American Indian shamanism to esoteric Judaism, this concept has dominated for millennia. As it has now become clear, western civilization is unique in history in its failure to recognize each human being as a subtle energy system in constant relationship to a vast sea of energies in the surrounding cosmos." - Dr. Edward Mann, sociologist

It is the western habit to define, generalize and compartmentalize, and health is no different.

Our oversight in deeming the body and mind separate is one that we must painstakingly unlearn and correct if we want to evolve with efficiency in our understanding of the human being.

What do illness and wellness mean for you? Perhaps the words "ease" and "dis-ease" paint a more apt picture? What is a mental illness? The definition and the concept of mental illness has changed and evolved in the west and continues to do so as we learn more about the mind, the body, and the connection between the two, and will continue to evolve as we reach out and embrace other worldviews of illness and wellness.

"When conscious life is characterized by one-sidedness and false attitudes, primordial healing images are activated - one might say instinctively - and come to light in the dreams of individuals and the visions of artists.... schizophrenia is a condition in which the dream takes the place of reality." - Carl Jung

It is undeniable that with the explosion of the bodymind concept in the 21st century, ideas regarding illness and health will transform as medical thinking becomes more patient-centered, organic, and holistic.

HEALING ONESELF

Anything and everything that has ever been conceived of has been conceived of from nothing. Theories, medical practices, martial arts, concepts... anything that is, wasn't, at one time. What does that mean? That means that individuals and groups daring enough to follow their instincts and their intuition are the people that have conceived of anything and everything that humanity has ever done or been.

Why should you be any different? Do you need someone else's permission before you make your decisions? Do you need society to agree first before coming to your own conclusions about you, your body, your mind, your soul and your potential?

Moshe Feldenkrais first injured his knee while playing soccer, and then re-injured it while working slippery submarine decks as a naval scientist during the Second World War. Feldenkrais was also a Judo teacher and had mostly completed his Doctor of Science, and as the prospect of surgery would leave him with a life-long limp, he decided to apply his knowledge of physics, engineering, and martial arts to an intensive study of his own movement habits.

Judo was a main influence on the Feldenkrais method, as it differentiates between rote exercise and attentive movement, "the methods of physical exercise in vogue... exert only the muscles without any other goal, and one needs much will to bind oneself unfailingly to one of these methods. Judo is very different, in that each movement has a specific goal which is reached after a precise and supple execution" – Feldenkrais

So essentially, Feldenkrais took his health into his own hands using his intuition, the knowledge at his disposal, the observation and awareness of his movement habits, and purposeful attention directed, at the moment, into and through his movements.

Feldenkrais also incorporated a variety of other schools of thought including "cybernetics," which is the "scientific study of control and communication in the animal and the machine," and is a transdisciplinary approach to exploring regulatory systems, their structures, constraints, and possibilities.

NOW WHAT

"The idea of victimage is a dreadful thing, a product of a safe middle-class perspective. What people who are not safe develop is tragic wisdom, the wisdom that embraces contradiction and seeks a sense of balance rather than going to extremes." - Gerald Vizenor

People all over the world and throughout the course of history have achieved the miraculous, pioneered new methods, and carved brave new frontiers in their decision to say "yes" to themselves, "yes" to their instincts and their calling, and "yes" to bold action and evolution. You are at the helm of your own dream, and no one can dream it but you. Gather yourself, gather your trophies and your understandings, gather your wounds and your tragedies, and use them to fuel your journey, to light your path, and to lift you higher. The journey through the dark night and to the dawn vanishes with the coming of the light, and in the glorious light of that proverbial morning, there only exists "right now," despite the lifetime that it took to reach that now.

You have the option, in every moment, to begin again at that moment. Every moment is new. Our bodies are ever in motion, changing over with the seasons, and intelligently evolving as we in turn service our own evolution. What if you could operate at your maximum potential?

What would that look like?

Throughout the course of this book, we've covered the waking dream, the mind-body connection, and the inherently holistic nature of health. We've explored the tales of neuroplasticity, and the experience of the miraculous, the power of an iron will be coupled with appropriate action, and the realm of the silent knowledge of the dreamer.

To be a dreamer, one must simply dare to dream.

ACTIVITIES, EXERCISES, AND IDEAS

Consider this portion of the book as a collection of topics discussed and subsequent supplementary or complementary activities.

Lucid Dreaming Technique:

WILD - "Wake-induced lucid dream"

- Lay down in your bed with your eyes closed
- Relax your body and mind completely (meditation is a good way to relax)
- Try to empty your mind
- Observe the state between waking and sleep, the "hypnagogia." Hypnagogic hallucinations are vivid, dream-like sensations that can be

potentially heard, seen, felt or smelled.

Let your mind wander around and notice anything that appears while in that state, and "follow" it.

- Begin creating the dream scene
- Begin visualizing in detail and explore your surroundings
- Stabilize yourself within the dream often by reminding yourself that you are dreaming

WAYS TO REACH YOUR "INNER WELL"

- Be authentic
- Be brave
- Trust your process
- Follow your signs
- Focus on yourself, don't worry about what others are thinking or doing
- Feel without judgment

- Find a breathwork session or practice on your own
- Journaling
- Solitude
- Appropriate Sleep

PROBIOTIC FOODS

- Yogurt
- Kefir
- Sauerkraut
- Tempeh
- Kimchi
- Miso
- Kombucha
- Pickles

20 WAYS TO CHANGE YOUR MIND

- Believe that you can
- Align your language to your desires

- Live in the moment
- Honor your body by listening to it
- Dialogue with your body
- Meditation
- Intended breathing
- Gratitude
- Repetition
- Purposeful movement and action
- Adequate Sleep
- Take the journey to the bottom of your well
- Lucid Dreaming
- The art of stalking
- Visualization
- Physical Activity
- Eat well
- Commit to reducing stress
- Be mindful of the effects of your environment and make appropriate changes; shake up your routine and improve your cognitive flexibility
- Be committed to keeping an open mind

BUILD BETTER HABITS

If you ever hope to change your mind for the better it is vital that you change the negative habits that make it easier for you to stay just the way you are. Broadly speaking, if you feel better you will have a better outlook on life.

Self-discipline: When it comes to habits that ensure that you get things done, one of the most impressive is the ability to exercise self-discipline at will. While, as with most habits, getting started can be difficult, the following tips are sure to make the process far more manageable.

Go all in: If you ever hope to improve your self-discipline, you are going to need to commit to the idea completely. Giving it a go while keeping one foot out the door will only lead you to shirking your discipline when it suits you, which is no true way to build a habit at all. Instead, you are going to want to commit to the idea fully and stay on top of your reactions to ensure that if they slip from the disciplined path, you are right there to ensure they get back in line. Making a decision to commit fully to the task at hand will go a long way towards silencing your inner critic. If you don't commit fully, you then ultimately run the risk of falling back into negative habits after months, or even years, and destroying all your hard work.

Keep an eye out for triggers: When it comes to getting in the habit of practicing self-discipline, it is critical that you take the time early on to consider the types of things that commonly trigger you to lose control. Getting a better handle on your triggers will make it easier to understand the underlying habits they prop up, which will make it easier for you to avoid the whole affair in the future. While you might not be able to think of any triggers right away, if you keep the topic on your mind, then as you go through your week you should notice things that are more likely to stray from the chosen path.

Removing triggers: Once you have managed to make a list of your triggers, the next thing you are going to want to do is everything in your power to ensure you remove them from your general line of sight until you have your habit of being self-disciplined down pat. While you will rarely be able to remove absolutely all the power a given trigger has, you should be able to lessen it significantly, with practice.

Understand excuses: When it comes to exercising your self-discipline, especially in a scenario that will require significant time and energy to completely successfully, it is completely natural for your mind to come up with excuses as to why it makes sense to take the easy way out, some of them might even be rather believable. If you find yourself routinely putting off tasks because you are afraid to overexert yourself, due to the commitment of those around you, because you have too much on your plate or due to external factors then you might need to take another look at those activities and determine how important they really are too you.

While it can be easy to believe excuses that your mind puts out, especially when crying off is the easier option when these situations arrive it is important to look into your heart and determine the true reason for the delay. What's more, however, once you look, you need to act on what you see, looking is easy, acting requires self-discipline and will become easier with time.

If you find it difficult to ignore the part of your mind that likes to generate excuses, making deals with it might work early on until you have flexed your self-discipline a little more. For example, if you are looking to get into shape and are having a hard time getting up to exercise every morning you can make it easier to ignore the possible excuses you might have by giving yourself some form of reward on the days you do exercise. Over time, and once you begin to see real results, you will find that you need the extra motivation less and the excuses will naturally fade away. Remember, perseverance is key.

If you still find yourself giving into excuses, make a concentrated effort to change the spin you are putting on the lies you are telling yourself. Instead of blaming failure on external factors, place the blame squarely at your own feet and tell yourself you are really crying off because you would rather do something easier, because you are scared, or even, simply because you are lazy. When other things don't work, confronting yourself with the blunt truth of the situation will often do the trick.

Change your routine: When building a lifestyle based on self-discipline the easiest was to assure that the new you hangs around for good, is to make sure that whatever you are trying to do (or not do as the case may be) you keep doing (or not) every day until new routines develop and eventually new habits form. Once new daily routines become a habit you can then focus your willpower in a new direction. When building a disciplined lifestyle, start with one facet of your life, turn new routines into habits and then move on to the next, before you know it you will be a whole new person.

Build new habits*:* When first asserting your will over your body, the conditions, the timing or anything else that you previously used as an excuse to keep you from doing whatever it is you knew needed doing, will make your new habit feel awkward and extremely difficult but the good news is that this is normal. This is simply what forming new habits feels like and there is no shortcut for it. The only thing that helps is knowing that it will get easier over time, just keep telling yourself that and it should help you through the rougher spots.

The easiest way to go about creating new habits is to mix up your routine to help avoid whatever triggers you may have that influence the negative behavior. Such urges are caused by the basal ganglia, a part of the brain that deals in memories, patterns, and emotions. The rational part of the brain which deals with making decisions is the prefrontal cortex and when an action becomes a habit it shifts from the second into the first. As such, changing your routine by inserting something new can trick the brain out of relying on the basal ganglia and force it to go back to using the prefrontal cortex.

Set the right goals: While childhood days may

have provided us with many ideas of the things we would like to accomplish in life, (becoming an astronaut, doctor, cowboy, to name a few select careers) the stages of late adolescence and adulthood tend to give the majority of people a drastically changed perspective. This, of course, isn't anyone's fault. The world just operates differently than most people think when they lack any informative worldly experience.

Again, this isn't any one person's fault, but rather it's everyone's. Almost every single person is trying to accomplish something, or get something, and so it stands to reason that at some point, two people with a common goal will try to get the same thing when there's really only enough for one of them – be it a job, or a romantic partner, or anything really. What the result of this tends to be, more often than not, is that the person with the least amount of dedication and self-discipline will fail while the other will attain the success that both originally pined for.

This isn't to suggest that a person – any person – couldn't accomplish these aforementioned goals if they really wanted to, it's just that usually, they are faced with the problem we discussed in the previous chapter: motivation. They have a starry-eyed view of the potential job or girl they've been crushing on, but they only see the end result. As if they time traveled, they picture themselves in the glorious career or with that beautiful woman, and they envision themselves accomplishing these things without really taking into account all of the dedication and focus that is a pre-requisite to get there in the first place.

Again, that initial spark could be seen as necessary, because without those starry-eyed young adults who dared to dream, our society would certainly collapse. We would find ourselves with a dangerous deficit of doctors, lawyers, and cowboys. But again, the only people who end up making it in those roles are the ones with self-discipline, because again, it isn't just about having the desire to do something; it's having the dedication to back that desire.

The reason that most people fail is that the goal they choose is too broad. For example, setting a goal of being rich doesn't take into account that it is really only an umbrella term that won't actually get you any closer to your ultimate goal. Instead of saying you want to be rich, setting a goal to be successful at a well-paying profession, for example, ensures the same end result while at the same time providing you with a number of guideposts along the path to ensure you can tell when you are moving forward and when you are only treading water.

Reduce social anxiety: Everyone experiences

some type of anxiety at one point in their lives. When you get up to speak in front of a group of people do you get a bit sweaty? Do you blush when you meet someone new and forget what to say? If your called on in class, do you feel nervous because everyone is looking at you? This is all situations that most people will face at some point and it is common to have the symptoms that I described. When you talk quiet and get nervous, do you blame it on your shyness? Shyness and social nervousness are quite normal. Overthinking and feeling nervous at a point in your life could be described as anxiety, but it does not mean that you suffer from social anxiety.

Imagine yourself being successful in social situations: As hard as it may be to believe, imagining yourself getting better at social interactions can actually have a measurable effect on your performance in the real world, but only if you go about doing it in the right way.

To get started, you are going to want to imagine yourself in a place that you are very familiar with, where you are likely to run into individuals that you do not know terribly well, if at all. When you are imagining this place, you are going to want to really visualize it. Picture every nook and cranny, think about the smells and the sounds you would experience and generally do everything you can in order to put yourself into that space as completely as possible.

Once you have a setting in mind, the next thing you are going to want to do is to put yourself into the space. However, you are not going to want to inhabit the space in your body as you would if you were really there, you will want to put yourself into the space as though you were viewing yourself in the third person. This is an important step as it will give you a buffer between yourself and all of those negative feelings which creates the unproductive feedback loop you are trying to avoid. You will notice how much easier it is to practice in this way if you try doing the exercise from the first-person view.

Once you are in the imaginary space and settled in a location you would be likely to inhabit in real life, the next thing you will want to do is to bring in someone you would like to have a conversation with. Remember, this should be someone you don't know at all or that you only know in passing. The specifics don't matter as long as you tend to be anxious when trying to speak with them in real life. It is important to start with just one person as opposed to a group when you are just starting out as you don't want the conversation to jump around too much for reasons that will soon become obvious.

With that done you will want to go ahead and imagine having a conversation with that person.

You are going to want to think through both sides of the conversations and make all the responses, both yours and the other person's as natural and realistic as possible. At first, you don't need to do anything more than play out the conversation as it would naturally occur and try to get a real flow going.

As you move through the conversation and come upon stumbling blocks that normally trip you up, pause the conversation, rewind and try again. Try out various responses and see how they land, then go back and try them again and again until you are happy with the alternatives you have chosen. Don't worry about how long it takes, or how many different tries are required until you get it right, this is just practice after all. Your goal with the conversation, and with every conversation in general, should be to eventually reach a point where you can address a topic that you are knowledgeable about to speak on at length or an aspect of the other person's life that you are genuinely interested to explore further.

Once you have successfully made it through an entire imagined dialogue, odds are you have now spent upwards of 30 minutes practicing genuine conversation, regardless of the fact that you were providing both sides of it. This exercise primes your brain to continue working on these types of problems in the background so that your brain will be more primed to utilize these types of neural pathways in the future. It will also help you to learn to push past the occasional awkward or suboptimal response and find a way to keep the conversation going under less than ideal conditions. Again, all without having to deal with difficulties inherent in practicing with a real person.

Once you make this type of practice a habit you will often find that conversations you've had in your head are now popping into your mind at random times without you even having to consciously bring them to the foreground. Once this happens you know you are on the right track. What's more, you will find that these types of positive mental conversations often replace the negative self-talk that likely popped up in this space beforehand.

The goal with this exercise is not to try and accurately predict what the other person is going to say in every situation, as that would be practically impossible. Rather, you should focus on making the conversation flow as smoothly as you can throughout. Once you have this activity down to a science, you are then ready to take the practice into the real world and start having productive conversations with strangers. While it is natural to apprehensive at this thought, armed with this exercise you will be astounded at how much easier to do so the experience becomes.

Ask for feedback: It is important to improve your interpersonal skills when you are dealing with social anxiety, simply because the way you see yourself in these situations so rarely matches the way that other people perceive you in the same situation. As such, while it will most definitely be difficult, one of the best ways to improve is to ask for feedback from those you interact with. At first you can go ahead and ask for feedback from those who you are comfortable interacting with, but eventually, you are going to need to go out on a limb and ask people you are less familiar with what they think of your social skills.

This is obviously going to be a difficult thing to do, but getting the perspective of relative strangers on your social strengths and weaknesses can be a big wakeup call when it comes to aligning the way you see yourself in these situations and the way others see you. The easiest way to go about doing so is to find a group of people whose opinion you don't especially care about and then go to town. Again, it is going to be awkward, but the results are likely to surprise you. Unless you are doing something extremely wrong, odds are they won't have anything but praise, or at least, anything negative to say. As a general rule, people are far more critical of themselves than they are of other people.

Spend time around those with well-defined social skills: Another useful exercise is to spend time around people who have already spent time honing their social skills. Finding a local Toastmaster's club is a good place to start. These meetings are filled with people who are gregarious and enjoy spending time with other people. While this may seem like the last place you would want to go, the opposite is actually true.

First, you will be able to watch these types of people in their natural habitat, which can help you learn the mannerisms that stick out to you and allow you to file them away for future use. Additionally, if you spend a lot of time around these types of people you will be surprised to learn that you will start to pick up their habits and mannerisms without consciously thinking about them.

What's more, as these are the types of people who are naturally outgoing, you will likely find that it is much easier to get into a conversation with any one of them without having to try very hard. Likewise, you will most likely find that they will naturally carry the conversation along without you having to necessarily contribute much at all. This means it is a great way to get into the habit of having regular conversations with people without having to worry about doing it wrong or letting your hang-ups about social situations come into play. After all, these types of meetings are all about talking to people and learning to do so more effectively two things that you are already trying to master.

Chapter 7: How To Improve Brain Health With Meditation

Mindfulness meditation is a type of meditation which focuses on being as aware of each moment as possible, thereby helping the consciousness to expand by forming a stronger connection with the present. Mindfulness meditation has a long history of practice as part of the Buddhist faith where it is revered for its ability to improve both mental happiness and physical well-being. This has been corroborated by research which shows that mindfulness meditation is a beneficial treatment for a variety of mental conditions. What's more, it has also been shown to be effective when treating conditions including anxiety, stress and drug addiction.

Practicing mindfulness is a skill and like all skills can be improved with practice. To practice mindful meditation, you simply try and retain as much focus on the current moment as possible with the help of repetitive breathing and the information being relayed by the senses. Studies have shown that practicing mindfulness for just 15 minutes per day can lead to measurable results when it comes to reducing stress and improving a sense of self. This is caused in no small part by the positive effects mindfulness has on emotional regulation, attention span, and body awareness. What's more, neuroimaging results show that practicing mindfulness also helps the mind process information more effectively.

Research shows that practicing mindfulness regularly can improve brain health as well as function and starting young will ensure your brain retains more volume as you age. Those who regularly practice mindfulness will also find they have a thicker hippocampus and as a result, have an easier time learning and retaining more information. They will also notice that the part of the amygdala which controls fear, anxiety, and stress is less active. With all of these physical changes to the brain is it any wonder that those who practice mindfulness report a general increase in well-being and mood?

Beyond the physical changes, regularly practicing mindfulness has been shown to decrease instances of participant's minds getting stuck in negative thought patterns while at the same time increasing focus. This should not come as a surprise given the fact that a recent Johns Hopkins study found that regularly practicing mindfulness meditation is equally effective at treating depression, ADD and anxiety.

In addition to the physical changes that you are likely to experience when meditating regularly, regularly practicing mindfulness meditation will also help you to more easily free your mind from any negative thought patterns you might otherwise find yourself getting stuck on making it easier to focus on the positive instead. Mindfulness meditation is so effective at this task that a recent study out of Johns Hopkins University actually found that it is just as effective at treating anxiety, depression and attention deficit disorder as many of the leading medications specifically designed to do the same thing. Another recent study also showed students preparing to take the Graduate Records Examination, the most common test to obtain admission into graduate school, who practiced mindfulness meditation regularly prior to testing scored approximately 10 percent better than their less mindful peers.

With so many physical and mental benefits, is it any wonder that mindfulness meditation is revered by Buddhists all around the world? The practice has its roots in a type of structured meditation called vipassana which, when translated, refers to a mental state that promotes living in the moment while still being aware of how the present and the future intertwine. Those who master vipassana are said to more fully understand the universe as a whole as well as their place in it.

In order to reach a state of vipassana, practitioners strive for what is known as the three marks of existence: impermanence, non-self, and dissatisfaction, which together are believed to bring unity to all living things. Non-self refers to the idea of understanding the boundaries between the self and the physical world with the understanding that coming to terms with these boundaries make it easier to fully grasp the intricacies of both. Meanwhile, dissatisfaction refers to the innate desire to seek satisfaction from fleeting experiences and the inevitable feeling that losing these things creates. This leads to the idea of importance as only by accepting the temporary nature of life can true happiness and inner peace be found.

If you thought mindfulness only had a benefit on yourself, you were wrong! Mindfulness is a practical and strategic way to benefit the world around you, as well! This amazing technique allows you the opportunity to improve your world through several different ways.

First, when you are mindful, you are much less likely to engage in arguments or conflict with other people. When you do choose to engage in a conflict, you will be much more rational about your approach, and the situation will likely diffuse quickly. If it doesn't, you will recognize that no benefit is being drawn from the experience, and you will remove yourself from the situation. Mindful people are generally much less emotionally charged in a negative format than those who are not mindful. They are more likely to be able to handle anger and stresses strategically, which means that even their "opponent" will end the situation in a more calm and rational way. This may simply diffuse one set of bad emotions, or it could trickle and encourage the other person to go learn about mindfulness and practice being more peaceful and calm in their own lives. You never know!

Additionally, the more we are in sync with the world around us, the more we are going to experience value from the world, and give value to the world. We are more likely to notice people who are struggling, so we can offer help. We are more likely to experience the highest joy we possibly can, which truly is contagious! Many other people who experience your intoxicating joy are going to turn around and experience some of their own as a result!

Finally, people who are mindful are generally a lot more considerate of the Earth itself. They tend to take better care of the world around them through many measures, including but not limited to recycling, not littering, helping clean up after others, taking care of plants and animals, and more! Doing all of this contributes to the healthy production and growth of the planet, which means that you are assisting it in thriving and maintaining its health!

GETTING STARTED TIPS

1. Find the time!

This is probably the most difficult part of meditation. Without time, we find it easy to make an excuse to skip meditating for the day. Don't. Meditation doesn't require leaving your home or any kind of special equipment. All you need is your time and some space.

2. Observe the moment

Mindfulness is not necessarily quieting the mind or finding an eternal state of calmness. The goal here is simple. We want to pay attention to the moment we are in without judging. When we judge a thought or something we may have done in the past, we tend to dwell on it. That isn't living in the moment and is not conducive to mindful meditation. While this is easier said than done, it is a crucial step to mindful meditation. With practice, it will be easy to achieve. Be mindful of the moment, of your senses and your surroundings.

3. Ignore those pesky judgments

Take notice of the times you are passing judgment while practicing. Make note of them and move on.

4. Always come back to observation and the present moment

It is easy for our minds to get lost in thought. Mindfulness meditation is the art of bringing yourself back to the moment, over and over, as many times as it takes. Don't get discouraged. In the beginning, you will find your mind wanders a lot. Reel it back in and keep moving forward.

5. Be kind

Even if your mind does happen to wander, and it will don't be hard on yourself. It happens. Acknowledge whatever thoughts pop up, put them to the side and get back on track.

As you can see, the basics are quite simple. These are the things you need to remember on a daily basis while you are practicing. What's important is that you find the time to implement the basics every day. Mastering the basics will make it much simpler for you to dive into the deeper aspects of mindful meditation, which we will be discussing a little later on. Before we move on, let's address some common questions people have about mindfulness. It's important to realize there is plenty of room for learning by trial and error. What works for one person may not necessarily work for you.

STARTING OFF STRONG

Choose a set time and stick to it: As with any burgeoning habit, it is important that you create a routine for your mindfulness meditation and stay with it if you hope for the practice to stick. It typically takes 30 days for a new habit to take root in your daily schedule which is why it is important to commit fully to practicing mindfulness meditation if you ever want it to become part of your routine. Due to its low impact nature, nothing external is required, it is very easy for many people to make excuses to get out of meditating, especially if their daily schedule is already filled to bursting.

If you find yourself always coming up with an excuse to get out of meditating in the moment, you may find the following piece of advice particularly useful. "Practice mindfulness meditation for fifteen minutes every day unless, of course, you are extremely busy in which case you should practice for thirty minutes instead." Don't let the outside world intrude on your potential for inner peace, find a time each day that works for you and stick with it no matter what; in a month's time, you will be glad you did.

Get started by focusing on the moment: While the ultimate goal of mindfulness meditation is to quiet the mind in an effort to find a state of internal calm despite the hustle and bustle of the outside world, many people find it difficult to achieve this state right out of the gate. Instead, you will likely find it easier to start to supplant any thoughts you might have by focusing all of your attention on the signals that your senses are relaying to you to the exclusion of everything else. While you might not feel as though you are receiving much data on the physical world, especially if you are practicing in a quiet, temperate space, the truth of the matter is that your brain naturally filters out approximately eighty percent of everything it receives, you just need to get in the habit of tapping into it.

Over time, you will learn to tune out the thoughts you have regarding your everyday routines and instead tap directly into whatever it is that is going on around you. When you do so, it is important to process the information that your senses are providing you, while at the same time making a conscious effort to not pass judgment or dig too deeply into anything that crosses your mind. Judging results in additional thoughts, one way or another, which tend to lead to even more thoughts, until it is practically impossible for you to focus on the task at hand.

Remember, when it comes to mindfulness meditation, the goal is to get as close as you can manage to the moment as possible, which means ignoring everything else that is going on, with the exception of what your senses are providing you. To reach this state, you will start by focusing on your breathing, especially on the way the air feels as it enters and exits your lungs, along with the way it smells and tastes.

Once you have narrowed your focus to only this band of information, the next thing you are going to want to do is to start expanding your observations to include the other sensations your body might be experiencing at the time as well. With the top level of your mind temporarily cleared of all your immediate thoughts, you can then focus on going deeper into yourself in search of the point where you mind is content not creating any new thoughts and simply exists in a relaxed, peaceful state.

Avoid your thoughts: When you first begin practicing mindfulness, it is perfectly natural for your mind to constantly fill with thoughts rushing to fill the void left by your previous actions. This typically occurs because you have trained yourself over the years, whether you realize it or not, to constantly move from one thought to the next, in hopes of solving the latest major crisis. This is, of course, practically the polar opposite of what you are striving for with mindfulness which is why it is only natural for you to expect a bit of an adjustment period.

Each time you feel these thoughts encroaching on your state of peace, you are going to want to approach them in the right way to maximize your time spent in meditation. First and foremost, you are going to want to approach these errant thoughts in a way that is completely devoid of judgment. You are going to want to avoid judging each thought and also refrain from judging yourself for having them. If you find yourself being sidetracked by a specific thought all you need to do is to mentally set it aside, center yourself on the task at hand and then continue on as before.

Again, it is perfectly natural for this to be a process that is more difficult than it sounds, but you may find it useful to think of your stream of consciousness as a stream of bubbles instead. Each thought is then enclosed in its own bubble, floating by you at a distance. You will then want to let each bubble pass you by and then disappear once it is out of sight. Alternatively, you may find it help to think of your stream of consciousness as an actual stream with a dam on either side. You then just need to close up both dams and the stream will dry up until you are finished with the current exercise.

Persevere: In order to ensure that practicing mindfulness becomes a habit that you can stick with, it is important to start practicing it with the correct mindset from the start. Specifically, you are going to want to keep a reasonable level of expectations to make the day to day practice more manageable. Keep in mind that it is perfectly natural for your mind to wander and seek out thoughts, even after you have been practicing for a prolonged period of time. Perseverance is the key here as only by pushing through the moments of distraction will you be able to find the success you seek. Ultimately you are seeking the level of mental blankness that occurs in the instant you have been asked a question but before the answer has come to your mind, reaching this state in a repeatable fashion is the key to mindfulness success.

EVERYDAY PRACTICES

You now know how to start your own mindfulness meditation practice, but you might be wondering how you can bring mindfulness into your everyday life. You also may not be interested in adopting a meditation routine, but you can still bring mindfulness into your life with a few simple actions.

As you have learned, humans have a tendency to go on autopilot, and this happens more often during the normal everyday tasks we have to do. These are the moments when you need to become more mindful. You don't have to clear your mind of everything, just become aware of what you are doing, and notice how it feels. Here are some activities where you can become more mindful.

Brush your teeth: When you brush your teeth you probably don't think about what you're doing. You've been doing it for years and it's not that hard. You stare at your reflection and focus more on how your skin looks than what you are doing. You may even have to run through your house with the toothbrush sticking out of your mouth.

Instead, start thinking about the texture and taste of the toothpaste and brush. Think of how the brush feels as you move it in your mouth. Think of how the floor feels under feet and your arm feels as it moves. Be mindful as you brush each of your teeth.

Wash dishes: Most people have a dishwasher now, but when you have to wash dishes by hand you moan as you approach the sink because of the menial task. You robotically scrub, rinse, and dry; over, and over again.

Instead, notice how it feels. Feel the water on your hands. Notice how the scrubber feels when you rub it against the dishes. Notice the difference between how the dirty dishes feel and the clean dishes feel.

Stand in line: There are lots of times where you will find yourself standing in line; the grocery store, shopping mall, DMV, wherever. You stand there, trying not to make eye contact, and groaning about the time that it's taking.

Instead, start looking at things, noticing them. Notice what the area really looks like. Look at the people around you, don't stare, they may take offense to that. Notice the smells, hopefully, they are pleasant. Take advantage of this moment to notice your surrounds, and to become more aware.

TIPS FOR MINDFULNESS SUCCESS

Stay Mindful of Your Actions: Too easily can we fall into a routine with too many bad routines in it. This can range anything from eating out too many times a week because you don't want to spend the time to cook to buying things on impulse because "hey, I've got a few extra dollars, so what's the harm?" While spontaneity is not intrinsically a bad thing, making decisions without considering them first can lead to a lack of understanding of the consequences.

When making a decision, one that may lead to a habit down the road, stay in the present and consider it. If you don't think about something as you're doing it (like buying fast food for dinner four times a week) it will become a habit without you even realizing it.

It works for good habits, too. Yes, waking up early on a Sunday morning to run in the cold is difficult, but if you stay in the present and remind yourself why you want to run, you'll find it easier to drag yourself out of bed. If you don't stay aware of the present, you'll find yourself pulling the covers over your head and falling right back to sleep.

Here's the kicker: Staying present and mindful of the situation or decision at hand is also a habit you have to force yourself to learn. So stay mindful of staying mindful and the rest just may fall into place.

Mindfulness of Your Thoughts: Mindfulness doesn't only affect how you view your actions; it also impacts how you think about your own thinking.

It's important to reflect on your thoughts throughout the day to understand your mind better and to change any harmful habits you may have. If you find yourself having negative thoughts about your habits, staying mindful of these thoughts will help you identify them and change them before it harms your progress toward a healthier you.

Your own thoughts about your habits or progress toward a healthier you are self-fulfilling prophesies. That is, whatever you believe about yourself will eventually happen. It's important to stay positive about your progress and understand that you may not see results right away, but if you keep telling yourself that you are getting healthier, the results will come after.

If, on the other hand, you keep telling yourself that you won't get any healthier because you can't see the results right away, you'll be more likely to give up and then definitely won't see any results.

Once you've moved past your doubts, you can begin to visualize your desired outcome. If you want to lose ten pounds, constantly visualize yourself losing ten pounds to help you continue your routine. If you visualize yourself staying how you are, you won't make progress.

Your mind is a powerful tool, don't let it work against you to prevent you from achieving your goals. Use it to your advantage.

Make a Schedule and Stick to it: What makes a habit a habit is the fact that you do it constantly without thinking having to force yourself into starting it. The only way to do this is to create a daily (or weekly) schedule and to force yourself to stick to it. The more you stick to your schedule, the easier it will be for you to stick to your schedule.

With technology today, creating and sticking to a schedule has never been easier. Smartphones are perfect for creating a calendar with alarms to go off up to a day before your activity is planned, so you can have plenty of time to pump yourself up.

Some of the best advice I've ever been given is also the simplest in theory: Set an alarm for when you need to do something. When the alarm goes off, do the task without hesitation.

If you have an alarm set for your daily tasks

(waking up, going to work, cleaning the kitchen, mowing the lawn, homework, the list is limitless) and do that task as soon as the alarm goes off, you'll have a much easier time creating habits and keeping them. Like all things, it's easy to make an excuse for not doing it at the time, but if you force yourself to get up and do whatever your alarm says to do, you won't have time to consider all of the countless excuses not to do it.

Stick to Your Habits: Habits aren't formed after only one or two days. For the first few weeks, you're going to have to force yourself to do whatever you want to make into a habit. On most of those days, you probably will hate it, but it gets easier after that first struggle.

On average, it takes about 21 days for an individual task to become a habit. That's three weeks of doing something that, honestly, you probably won't enjoy too much to reach a point where you can stand it. It's hard and can suck, but if you remind yourself of your reasons for doing it and stick to the schedule, you'll get to the point where you can't live without your habit.

Once you form a habit over those three weeks of pushing yourself, you may not fully appreciate how easy the task becomes. You may still hate running in the morning because of how hard it can be to wake up early. But once you get a habit locked in place, you won't be able to go a day or two without it without feeling the negative effects of not doing it.

It comes down to one thing to form a habit: Tenacity. It's all about pushing yourself to continue a task even when it's difficult to make it a habit in your everyday life. The aforementioned tips and tricks are designed specifically to make being tenacious easier and, therefore, making forming and keeping habits easier.

Make it an intention to improve your overall consciousness: If you have the motivation to improve your state of awareness, you are already on the right path to success. Having an intention alone will help you to home in on finding new ways to raise your self-consciousness continuously.

Be truthful: When you speak the truth, you immediately raise your consciousness levels. People tend to be dishonest during times when their consciousness is at lower levels. Individuals with higher levels of awareness tend not to lie because they respect themselves and wish to be true to their overall being. This aids in assisting all sorts of relationships within our society to become more conscious in nature.

Live for your purpose: While you never know when or how you will discover your purpose in life, once you do find a spark, take care of that little glow of light so that in time it will help you grow a fire. Once you harbor your sense of worth and purpose, you are then more capable of sharing it with the world, which will continue to help you raise your sense of awareness.

Be conscious of your decisions: When you are avidly in control of the decisions you make, you unknowingly activate neural pathways within the brain that aid in promoting inner peace, calmness, and self-control. If you are allowing others to take control of the decisions you make, you are not fully conscious and are less able to take responsibility for your actions.

Be open-minded: Always being open-minded is an aspect of the overall process of becoming more aware. If you fail to accept the diversity that resides naturally in our societies, you are only inhibiting the world from giving you loads of unique opportunities. It also keeps you from being more aware of life. Being open-minded means being attentive enough to go out and try new things, such a new cuisine, exercise routines, and other such things.

Seek out higher intelligence: There are many opportunities that one can partake in to become more intelligent. No one is keenly smart in all aspects of their life, and there is always something new to learn. While some people are more emotionally intelligent, others have a higher I.Q. level. Enhancing any form of intelligence means that you must be consciously aware in order to pave the journey to expand your horizons.

Conclusion

Thanks for making it through to the end of **Rewire Your Mind**: *How To Change Your Mind To Live A Successful And Positive Life On Your Own Terms*, let's hope it was informative and able to provide you with all of the tools you need to achieve your goals, whatever it is that they may be. Just because you've finished this book doesn't mean there is nothing left to learn on the topic, and expanding your horizons is the only way to find the mastery you seek.

Now that you have made it to the end of this book, you hopefully have an understanding of how to get started improving your mindset for the better, as well as a strategy or two, or three, that you are anxious to try for the first time. Before you go ahead and start giving it your all, however, it is important that you have realistic expectations as to the level of success you should expect in the near future.

While it is perfectly true that some people experience serious success right out of the gate, it is an unfortunate fact of life that they are the exception rather than the rule. What this means is that you should expect to experience something of a learning curve, especially when you are first figuring out what works for you. This is perfectly normal, however, and if you persevere you will come out the other side better because of it. Instead of getting your hopes up to an unrealistic degree, you should think of your time spent working on yourself as a marathon rather than a sprint which means that slow and steady will win the race every single time.

Especially when you are first transitioning from fixed mindset to growth mindset, you might find yourself putting a lot of emphasis on speed. "How fast can I do this?" might be your thought, often. This is because fixed mindset people are more focused on the rewards than the process. If you want to have a growth mindset, you need to transition your emphasis from the reward to the growth itself. Focus on the growth. Instead, ask yourself "How much can I grow from this?" and "How can I maximize my growth from this experience?" When you do this, you successfully transition to a growth mindset and place your emphasis on the growth.

Reflection is the best time to recognize whether you are making the progress you want to make, or if you are not growing as much as you could be. Take time to reflect so that you can see how far you have come and where your strengths and weaknesses are. This is a great time to identify any fixed mindset patterns that are still existing within' you and work towards healing them.

The "ideal image" of whom people are supposed to be and what we are supposed to be like tends to be where fixed mindset and perfectionism are rooted. If you want to abandon fixed mindset and foster growth mindset, you need to be willing to abandon the ideal image and pay attention to whom you want to be and what your ideal sense of self is. Then, work towards the growth that will get you there.

Goals tend to be major motivators for growth mindset folks. If you are willing to embrace your goals and work towards them on a regular basis, then you can almost guarantee that you are going to learn new things. If you are not, you are not setting your goals high enough. You should be willing to set new goals and work towards them consistently. Every time you reach a goal, set a new one. Have a few on the go at any given time so that you consistently have something to work towards. If you do not know how many is enough, focus on working on one short-term goal, one mid-term goal and one long-term goal at all times. This ensures that you are regularly focusing on learning new things of all sizes.

While over time it will become easier and easier to have the discipline to follow through on a goal, no matter the roadblocks in your path, it is important to understand the distractions that stand between you and improving yourself. The advent of smartphones has trained many people to adhere to an instant gratification mindset which states that anything that is worth doing is going to be immediately entertaining. This is completely counterintuitive to a self-disciplined mindset which is focused on deferring short term rewards for greater rewards at a later date.

It is important to understand that when it comes to self-discipline you are not a beautiful and unique snowflake, everyone else, even those who are extremely successful, has all of the same urges and tasks fighting for their time. What makes them extremely successful is their ability to dedicate themselves to the task at hand when the situation requires it. Don't wait for inspiration to strike or let modern electronic distractions get in your way, have the self-discipline to put your goals first.

If you find yourself losing the battle with negative thinking and considering abandoning all you have fought so hard to achieve, it is important to remember how long it took for you to reach the point you currently find yourself in. This pattern indicates that it is irrational for you to expect changes to come about any more swiftly. During these times it is important to understand how far you have really come and the positive improvements you have already seen. Remember, the only time you should be cross without yourself about a lack of positive progress is when you let that one instant be the catalyst to let bad habits start to sneak back in.

www.ingramcontent.com/pod-product-compliance
Lightning Source LLC
Chambersburg PA
CBHW031149020426
42333CB00013B/573